Number properties

Number machines

A1 In 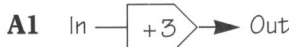 Out

    (a) In 4 gives an *out* of 7
    (b) In 0 gives an *out* of 3 (0 + 3 = 3)
    (c) 63 + 3 = 66

A2   (a) $7 \times 2 = 14$ (b) $0 \times 2 = 0$
    (c) $20 \times 2 = 40$

A3 In  Out

    (a) When you put in 3, you get 12 out.
    (b)   0 —[+3]— 3 —[×2]— 6
    (c) $5 + 3 = 8$ and $8 \times 2 = \mathbf{16}$
    (d) This one asks what you need to put in
        to have 8 come out.
        If you got an answer of 22 try again.
        The correct answer is 1 – how did you
        find it?

A4 In  Out

    (a) $6 - 3 = 3$ and $3 \times 2 = 6$ and $6 + 5 = 11$.
    (b) When you feed in 10 you get 19 out.
    (c) $3 - 3 = 0, 0 \times 2 = 0$ and $0 + 5 = 5$.

A5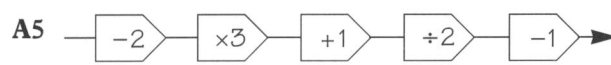

    (a) There are 5 machines in the chain.
    (b) $5 - 2 = 3, 3 \times 3 = 9, 9 + 1 = 10,$
        $10 \div 2 = 5, 5 - 1 = \mathbf{4}$
    (c) When you feed in 3, 1 comes out.
    (d) If you put in 7, you get 7 out!

A6 Here are two different ways of getting from
    2 to 6.

    2 —[+4]— 6 **and** 2 —[×3]— 6

A7 Each of these chains give 12 when 3 is
    put in.

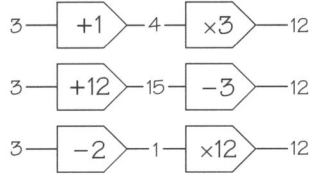

    3 —[+1]— 4 —[×3]— 12
    3 —[+12]— 15 —[−3]— 12
    3 —[−2]— 1 —[×12]— 12

A8, A9 and **A10** There are many answers to
    these questions.
    Show your answers to your teacher.

B1   6 —[×8]— [−5]— 43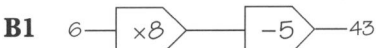

With questions like these it is probably a
help to work backwards.

B2   42 —[−2]— [÷5]— 8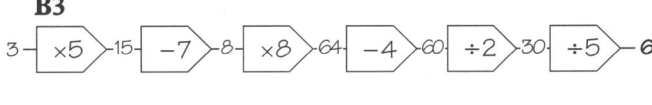

B3

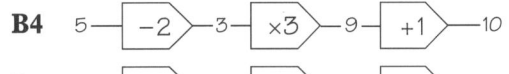

3 —[×5]— 15 —[−7]— 8 —[×8]— 64 —[−4]— 60 —[÷2]— 30 —[÷5]— 6

B4   5 —[−2]— 3 —[×3]— 9 —[+1]— 10

B5   7 —[−3]— 4 —[÷2]— 2 —[+5]— 7

C1 Here is the completed **input-output table**
    for this machine.

    In 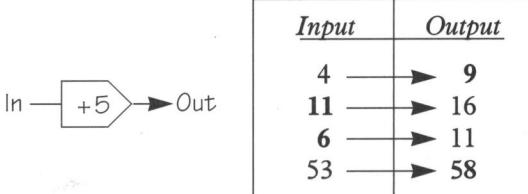 Out

Input	Output
4 →	7
6 →	9
9 →	12
23 →	26
17 →	20

C2 This rule gives this table.

In —[+5]— Out

Input	Output
4 →	9
11 →	16
6 →	11
53 →	58

C3 This is the only rule which fits this table.

In —[×2]— Out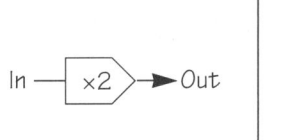

Input	Output
3 →	6
4 →	8
10 →	20
31 →	62

1

C4 This is the rule which fits this table.

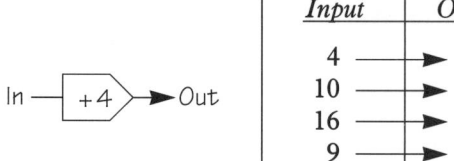

Input	Output
4	8
10	14
16	20
9	13

C5 This table fits this rule.

Input	Output
5	2
7	4
6	3
10	7

In — -3 → Out

C6 This rule fits this table.

In — $\times 4$ → Out

Input	Output
2	8
5	20
3	12
10	40

C7 Show your table to your teacher.

D1

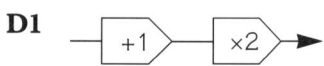

The completed table for this rule is:

Input	Output
6	14
3	8
9	20
49	100

D2

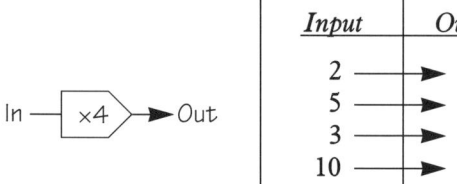

These two machines are the ones which fit this table.

Input	Output
3	9
7	17
10	23
27	57

D3

$\times 3$ — $+1$ →

These two machines fit this table.

Input	Output
2	7
4	13
10	31
16	49

Doing and undoing

A1 (a)

Real length in cm	Length in glass in cm
2	6
3	9
1	3
4	12
10	30
5	15
20	60
6	18

(b) The lengths in the glass are three times the real lengths.
So to find the real lengths you need to divide by three.
(Look at the table if you are not sure about this.)

A2 (a)

Real length in cm	Length in $\times 5$ glass in cm
3	15
4	20
6	30
1	5
0·3	1·5

(b) To **undo** 'multiply by 5' you need to divide by 5.

A3 (a) To **undo** 'multiply by 7', you divide by 7.
(b) Divide by 20 will **undo** 'multiply by 20.'
(c) Dividing by 43 will **undo** 'multiplying by 43.'
(d) To **undo** 'divide by 5' you need to multiply by 5.

A4 (a)

Real length in cm	Length in ÷3 glass in cm
12	4
6	2
15	5
30	10
18	6

(b) To **undo** 'divide by 3' you need to multiply by 3.

A5 (a) When the altimeter reads 100 m the real height is 400 m.

(b) If the altimeter reads 300 m, the real height is 1200 m.

(c) To find the real height you multiply the altimeter reading by 4.

A6 (a) Multiplying by 4 undoes 'divide by 4.'

(b) To undo 'divide by 6', multiply by 6.

(c) Divide by 8 to undo 'multiply by 8'.

(d) Multiplying by 8 will undo 'divide by 8.'

(e) To undo 'multiply by 100', divide by 100.

(f) You must multiply by 50 to undo 'divide by 50'.

B1 ? stands for 43.

B2 (a) $? \times 15 = 210$, so $? = 210 \div 15 = 14$

(b) $? \times 23 = 621$, so $? = 621 \div 23 = 27$

(c) $? = 7$ (d) $? = 83$ (e) $? = 29$

B3 (a) $? = 24$ $(? \times 15 = 360)$

(b) $? = 11$ $(? \times 21 = 231)$

B4 In the puzzle $? \div 6 = 21$, so $? = 126$

B5 (a) 598 $(? \div 13 = 46$, so $? = 46 \times 13)$

or

? —[÷13]► 46

?◄—[×13]— 46

(b) 551 $(? \div 29 = 19$, so $? = 19 \times 29)$

(c) 14 013 $(? \div 81 = 173, ? = 173 \times 81)$

(d) 13 769 $(? \div 49 = 281, ? = 281 \times 49)$

B6 (a)

? —[÷23]► 16

?◄—[×23]— 16

$16 \times 23 = ?$ so $368 = ?$ or $? = 368$

(b) 7688 $(? = 124 \times 62)$

B7 Don't forget to check that your answers fit the puzzle.

(a) 17 $(238 \div 14)$ (b) 3332 (238×14)

(c) 4966 (191×26) (d) 22 $(2882 \div 131)$

(e) 17 $(289 \div 17)$ (f) 1083 (57×19)

C1 (a) 83 kg   (b) 36 kg   (c) 20 kg

(d) 152 kg

(e) Subtracting 10 from what the machine says gives the real weight.

C2 (a) 54 kg   (b) 82 kg   (c) 141 kg

(d) 59 kg   (e) Add 10.

C3 (a) 'Subtract 7' will undo 'add 7'.

(b) 'Add 19' will undo 'subtract 19'.

(c) To undo 'subtract 34' you 'add 34'.

(d) 'Subtract 63' undoes 'add 63'.

D1 $? = 185$

D2 (a) $? = 46$ (b) $? = 171$ (c) $? = 26$

(d) $? = 48$ (e) $? = 97$ (f) $? = 41$

D3 (a) $? = 16$ (b) $? = 110$ (c) $? = 1044$

(d) $? = 17$ (e) $? = 629$ (f) $? = 3886$

D4

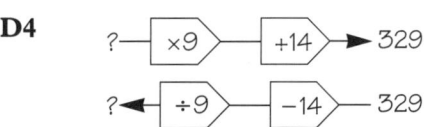

$329 - 14 = 315$, $315 \div 9 = 35$, so $? = 35$

Check by putting 35 through the original machine chain.

$35 \times 9 = 315$, $315 + 14 = 329$, it checks!

E1 He started with 15.

E2 I key in a number

I multiply it by 19.   I add 28   The answer is 503

? —[×19]—[+28]► 503

?◄—[÷19]—[−28]— 503

$503 - 28 = 475$, $475 \div 19 = 25$

So the number I started with was 25.

E3 (a) $? = 34$ (b) $? = 48$ (c) $? = 1440$

(d) $? = 14$ (e) $? = 8$ (f) $? = 191$

E4 (a) 5   (b) 18   (c) 32

E5 You need to feed in 135 volts to get 9 volts out.

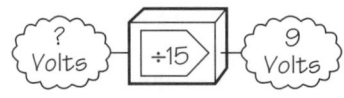

E6 When 160 volts are fed in, 4000 volts come out.

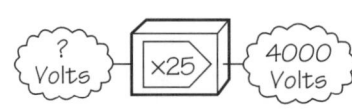

E7 It may help to look at the machine chain.

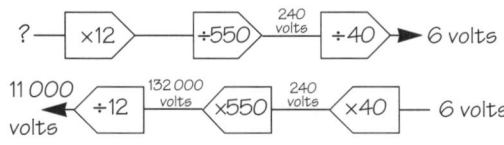

(a) 132 000 volts　(b) 11 000 volts

F1

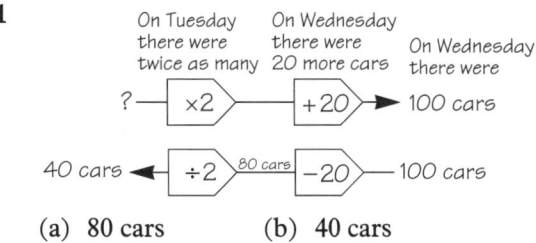

(a) 80 cars　(b) 40 cars

F2

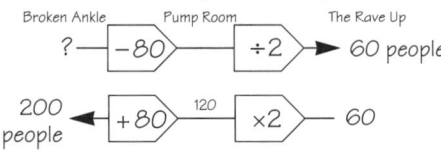

David caught 18 fish on Friday.

F3 Follow this through carefully.

Broken Ankle ─ ?──[−80]── Pump Room ──[÷2]── The Rave Up ── 60 people

200 people ──[+80]── 120 ──[×2]── 60

There were 200 people at the Broken Ankle Disco.

F4 45 sweets

F5 (a) 12 absent　(b) 60 absent

Number relationships

A1 Go back and check any of these which you got wrong.
(a) 18　(b) 28　(c) 16　(d) 45
(e) 32　(f) 21　(g) 80　(h) 42

A2 (a) $3 \times 8 = 24$　(b) $7 \times 5 = 35$
(c) $48 = 6 \times 8$　(d) $24 = 4 \times \mathbf{6}$
(e) $36 = 6 \times 6$　(f) $36 = \mathbf{4} \times 9$

You may find this multiplication table useful.

×	1	2	3	4	5	6	7	8	9	10
1	1	2	3	4	5	6	7	8	9	10
2	2	4	6	8	10	12	14	16	18	20
3	3	6	9	12	15	18	21	24	27	30
4	4	8	12	16	20	24	28	32	36	40
5	5	10	15	20	25	30	35	40	45	50
6	6	12	18	24	30	36	42	48	54	60
7	7	14	21	28	35	42	49	56	63	70
8	8	16	24	32	40	48	56	64	72	80
9	9	18	27	36	45	54	63	72	81	90
10	10	20	30	40	50	60	70	80	90	100

B1 15 squares are uncovered.

B2 28 squares are uncovered.

B3 25 squares are uncovered.

B4 54 squares are uncovered.

B5 Here are four ways of getting 6 in the corner of the multiplication table.
$6 = 1 \times 6$,　$6 = 6 \times 1$,　$6 = 2 \times 3$,　$6 = 3 \times 2$

B6 $16 = 4 \times 4$,　$16 = 8 \times 2$,　$16 = 2 \times 8$

B7 Here are all the ways you can get 24 in the corner of the multiplication table.
$24 = 4 \times 6$,　　$24 = 6 \times 4$,　　$24 = 3 \times 8$,
$24 = 8 \times 3$

B8 (a) These are factor pairs of 14:
1, 14　　14, 1　　2, 7　　7, 2
(b) These are factor pairs of 20:
1, 20　　20, 1　　2, 10　　10, 2
4, 5　　　5, 4
(c) 1, 30　　30, 1　　2, 15　　15, 2
3, 10　　10, 3　　5, 6　　6, 5

(d) 1, 36   36, 1   2, 18   18, 2
     3, 12   12, 3   4, 9   9, 4   6, 6
(e) 1, 25   25, 1   5, 5
(f) 1, 42   42, 1   2, 21   21, 2
     3, 14   14, 3   6, 7   7, 6

B9 (a)   (b)

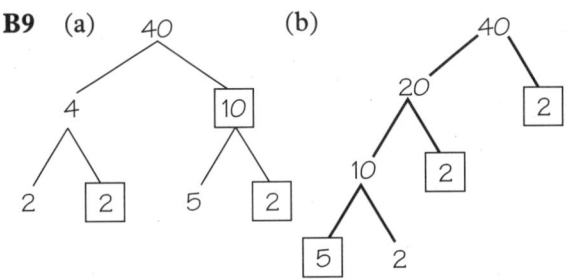

B10 (a)   (b)

(c)   (d)

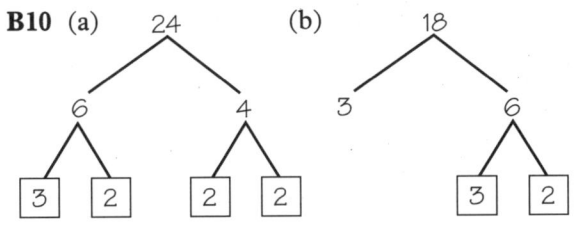

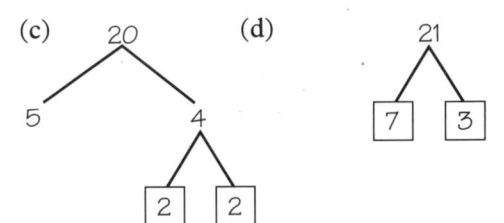

B11

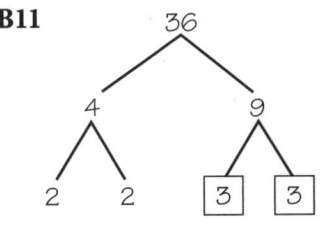

B12 $7 \times 2 \times 2 = 28$ $(7 \times 2 = \mathbf{14}, \mathbf{14} \times 2 = \mathbf{28})$

B13

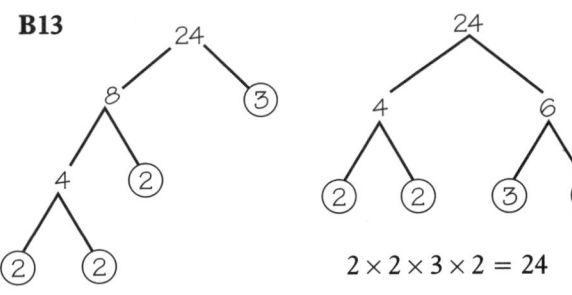

$2 \times 2 \times 2 \times 3 = 24$

$2 \times 2 \times 3 \times 2 = 24$

B14 There are many ways of drawing factor trees for 60.
Here is one of them.
Check that your own end numbers are the same.

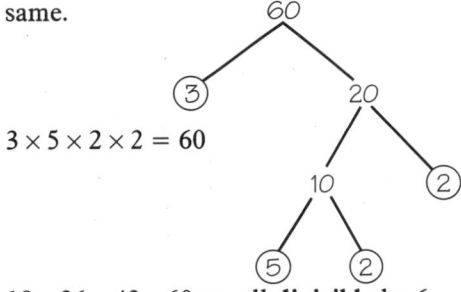

$3 \times 5 \times 2 \times 2 = 60$

C1 18   36   42   60 are all **divisible** by 6.

C2 The pattern 6 12 18 24 30 36 42 48 54 60 carries on 66, 72, 78.

C3 The numbers in row three are 3 6 9 12 15 18 21 24 27 30 **33 36** 39 42

C4

	Is 12 divisible by	
(a)	2	yes ($12 \div 2 = 6$)
(b)	3	yes ($12 \div 3 = 4$)
(c)	4	yes ($12 \div 4 = 3$)
(d)	5	no
(e)	6	yes ($12 \div 6 = 2$)
(f)	7	no

(g) – (j) Twelve is not divisible by 8, 9, 10 or 11.
(k) Twelve is divisible by 12 ($12 \div 12 = 1$, 12 divides exactly into 12).

C5 (a) 8 is divisible by 1, 2, 4 and 8.
    (b) 9 is divisible by 1, 3 and 9.
    (c) 10 is divisible by 1, 2, 5 and 10.
    (d) 12 is divisible by 1, 2, 3, 4, 6 and 12.
    (e) 18 is divisible by 1, 2, 3, 6, 9 and 18.
    (f) 20 is divisible by 1, 2, 4, 5, 10 and 20.
    (g) 7 is divisible by 1 and 7 only.

C6 (a) Yes   (b) Yes   (c) Yes   (d) No

C7 (a) Yes   (b) No

C8 There are several different ways of putting 18 stars into equal rows.

★ ★ ★ ★ ★ ★ ★ ★ ★ two rows of nine
★ ★ ★ ★ ★ ★ ★ ★ ★

★ ★
★ ★
★ ★ nine rows of two
★ ★
★ ★
★ ★
★ ★
★ ★
★ ★

You could also have 1 row of 18 and 18 rows of 1.

C9 Here are all the ways of putting 24 stars in equal rows:
3 rows of 8,   8 rows of 3,   6 rows of 4,
4 rows of 6,   2 rows of 12,   12 rows of 2,
1 row of 24,   24 rows of 1

C10 See page 10 of the booklet (13 is only divisible by 1 and itself).

C11 11 is a prime number. It is only divisible by 1 and itself.

C12 5, 11, 19 and 29 are prime numbers.

C13   T S I H D S U A O R Y K
or   T R A S H S Y O U D I K
or   T A H Y U I R S S O D K
or   T U S A I O H R D Y S K

C14 (a) THIS IS EASY
(b) FRED NOT TO BE TRUSTED
(c) ENEMY SHIPS SIGHTED OFF KENT
(d) TWENTY EIGHT IS DIVISIBLE BY FOUR

D1–D6 Show your worksheet to your teacher.

E1 (a) $4 + \boxed{5 \times 2} = 14$ (b) $\boxed{4 + 5} \times 2 = 18$

(c) $10 - \boxed{2 \times 3} = 4$ (d) $\boxed{10 - 2} \times 3 = 24$

(e) $\boxed{12 \div 3} + 1 = 5$ (f) $12 \div \boxed{3 + 1} = 3$

E2 (a) $(6 - 2) \times 4 = 4 \times 4 = 16$
(b) $3 + (2 \times 6) = 3 + 12 = 15$
(c) $20 - (6 \times 3) = 20 - 18 = 2$
(d) $(20 \div 4) + 1 = 5 + 1 = 6$

(e) $20 \div (4 + 1) = 20 \div 5 = 4$
(f) $7 - (5 - 3) = 7 - 2 = 5$
(g) $18 \div (6 \div 3) = 18 \div 2 = 9$
(h) $(18 \div 6) \div 3 = 3 \div 3 = 1$
(i) $3 \times (6 \div 2) = 3 \times 3 = 9$
(j) $(3 \times 6) \div 2 = 18 \div 2 = 9$
(k) $20 - (9 - 4) = 20 - 5 = 15$
(l) $(20 - 9) - 4 = 11 - 4 = 7$

E3 (a) There are many possibilities. Show your teacher if you are not sure.
(b) $(12 \div 3) + 6 = 10$ or $12 - (6 \div 3) = 10$ or $6 + (12 \div 3) = 10$

E4 (a) $(12 + 3) - 6$ or $(3 + 12) - 6$ or $(12 - 6) + 3$ or $3 + (12 - 6)$ make 9.
(b) $(12 \div 3) \times 6$ or $6 \times (12 \div 3)$ or $(12 \times 6) \div 3$ or $(6 \times 12) \div 3$ make 24.
(c) $(12 - 3) + 6$ or $6 + (12 - 3)$ or $(12 + 6) - 3$ or $(6 + 12) - 3$ make 15.
(d) $(6 \div 3) + 12$ or $12 + (6 \div 3)$ give 14.
(e) $(12 - 6) \div 3$ makes 2.
(f) $(12 - 3) \times 6$ or $6 \times (12 - 3)$ make 54.

The factor pair puzzles

This is how you could cover the first board with factor pairs.

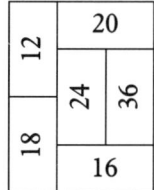

For the second board there are three ways to cover it with factor pairs.

24	12	20
18	16	36

24		20
12		
18	16	36

24		20
		12
18	16	36

Squaring and cubing

Number	Difference of squares	Number	Difference of squares
1	$1^2 - 0^2$	11	$6^2 - 5^2$
2		12	$4^2 - 2^2$
3	$2^2 - 1^2$	13	$7^2 - 6^2$
4	$2^2 - 0^2$	14	
5	$3^2 - 2^2$	15	$8^2 - 7^2$ or $4^2 - 1^2$
6		16	$4^2 - 0^2$ or $5^2 - 3^2$
7	$4^2 - 3^2$	17	$9^2 - 8^2$
8	$3^2 - 1^2$	18	
9	$5^2 - 4^2$	19	$10^2 - 9^2$
10		20	$6^2 - 4^2$

A1 Shena has laid (a) 9 tiles (b) 25 tiles.

A2 She has laid (a) 36 tiles (b) 49 tiles.

A3 (a) $5 \times 5 = 25$ (b) $1 \times 1 = 1$
(c) $10 \times 10 = 100$ (d) $0 \times 0 = 0$

A4

I think of a number, then square it. The answer is 81.

We are looking for a number which when it is squared gives an answer of 81.
$8 \times 8 = 64$ is too small, $10 \times 10 = 100$ is too large, but $9 \times 9 = 81$ is spot on!
Dave's starting number was 9.

A5 (a) $3^2 = 3 \times 3 = 9$ (b) $5^2 = 5 \times 5 = 25$
(c) $1^2 = 1 \times 1 = 1$
(d) $10^2 = 10 \times 10 = 100$

A6

Ten squared is one hundred. So twenty squared must be equal to two hundred.

Laura is not right.
$10^2 = 100$, but $20^2 = 20 \times 20 = 400$!

A7 It's probably clearer if you make a table of
▲ your results.
(Remember zero is a number.)

Challenge

The sequence of numbers which cannot be made from the difference of two squares (see the table for **A7**) is:
  2   6   10   14   18 . . .
Can you write a machine chain rule for this?

A8 Daniel and Pat think that any number can be made by adding at the most three squared whole numbers. They are wrong because 7, 15, 23, 31, . . . all need four square whole numbers, $7 = 1^2 + 1^2 + 1^2 + 2^2$ and $15 = 3^2 + 2^2 + 1^2 + 1^2$.
In fact, any number can be written using at the most four whole numbers squared.

A9 Your own method of squaring numbers on your calculator.

A10 (a) $1111^2 = 1234321$
(b) $111 \cdot 1^2 = 12343 \cdot 21$
(c) $11 \cdot 11^2 = 123 \cdot 4321$
(d) $1 \cdot 111^2 = 1 \cdot 234321$
(e) In each case there are the same digits, but although the numbers to be squared move one place to the right, the answers move two places.

A11

The square of any number is always larger than the number itself.

What about 1^2, $0 \cdot 1^2$ or $0 \cdot 9^2$?

A12 (a) A number and its square are never equal. What about 0^2, 1^2?

(b) Two will divide exactly into any whole number squared, in other words are square numbers always even?
What about 1^2, 3^2, 5^2, 7^2, … ?

(c) Squares of whole numbers never end in a 2 or 3.
We only need to look at the squares of the digits 0 to 9. Why?
$0^2 = 0$ $1^2 = 1$, $2^2 = 4$, $3^2 = 9$,
$4^2 = 16$, $5^2 = 25$ $6^2 = 36$,
$7^2 = 49$, $8^2 = 64$, $9^2 = 81$.

(d) If you square a number multiplied by 3 the answer is 9 times the square of the original number.
Let's try a few examples.

Number	(Number)^2	3 × Number	(3 × Number)^2
1	1	3	9
2	4	6	36
3	9	9	81

It looks as if the rule works, but we can only prove that it works **all the time** by using some algebra. We could do this by adding another line to the table.

n	n^2	$3n$	$(3n)^2 = 9n^2$

B1 (a) $3^2 = 9$ (b) $1^2 = 1$ (c) $0^2 = 0$
(d) $12^2 = 144$

B2 (a) The square root of 100 is 10 ($10^2 = 100$).
▲ (b) The square root of 121 is 11 ($11^2 = 121$).

B3 Here are all the square numbers between 5 and 50. (These are square numbers *not* numbers squared!)
9   16   25   36   49

B4

We had to find the square root of 20.
4^2 is 16 and 5^2 is 25,
so the square root of 20 must be between 4 and 5.

Trial	(Trial)^2		
4·5	$4·5^2 = 20·25$	too large	
4·4	$4·4^2 = 19·36$	too small	
4·45	$4·45^2 = 19·8025$	too small	
4·46	$4·46^2 = 19·8916$	too small	

(a) Looking at their table, a sensible next trial would be 4·47.

(b) How close was your final trial number?

B5 When you use trial and improvement it usually helps to use a table. It helps to see which trial values to choose. You may have found the answer quicker than here!
We are looking for a number, which gives 6·25 when it's squared.

Number (trial)	(Number)^2	
2·1	4·41	too small
2·8	7·84	too large
2·6	6·76	very close
2·5	6·25	spot on!

B6 The square roots of all these whole numbers end in a nine.
$\sqrt{81} = 9$ $\sqrt{361} = 19$ $\sqrt{841} = 29$ and so on

You might have saved time by working backwards $9^2 = 81$, $19^2 = 361$ etc.

B7 (1) $\sqrt{100} = 10$ is correct.
▲
(2) $\sqrt{4^2} = 4$ is correct.

(3) $\sqrt{1} = \frac{1}{2}$ (0·5) is wrong.
$\sqrt{1} = 1$ since $1^2 = 1$.

(4) $7 = \sqrt{3^2 + 4^2}$ is wrong.
$\sqrt{3^2 + 4^2} = \sqrt{9 + 16} = \sqrt{25} = 5$.

(5) $1 = \sqrt{2}$ is wrong. $1^2 = 1$, not 2.

(6) $\sqrt{12^2 + 5^2} = 13$ is correct since
$\sqrt{12^2 + 5^2} = \sqrt{144 + 25} = \sqrt{169} = 13$.

B8 (a) $\sqrt{196} = 14$ (b) $\sqrt{484} = 22$ (c) $\sqrt{961} = 31$
$14^2 = 196$ $22^2 = 484$ $31^2 = 961$

C1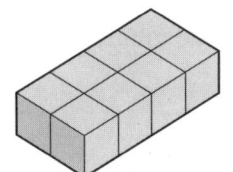

(a) 8 cubes (b) $3 \times 8 = 24$ cubes.

C2 (a) 4 layers
(b) 4 cubes in each layer
(c) $4 \times 4 = 16$ cubes in total

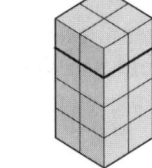

C3 (a) $2 \times 2 \times 4 = 16$ cubes
(b) $2 \times 2 \times 2 = 8$ cubes
(c) $3 \times 4 \times 2 = 24$ cubes
(d) $4 \times 2 \times 2 = 16$ cubes

C4

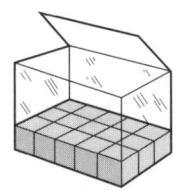

$5 \times 3 \times 4 = 60$ cubes

C5 (a) $2 \times 3 \times 5 = 30$ cubes
(b) $5 \times 3 \times 3 = 45$ cubes
(c) $4 \times 4 \times 4 = 64$ cubes

D1 Your own answer.

D2 (a) 2^3 (or two cubed) $= 2 \times 2 \times 2 = 8$
(b) $3^3 = 3 \times 3 \times 3 = 27$
(c) $1^3 = 1 \times 1 \times 1 = 1$
(so one cubed is one)
(d) Five cubed is $5 \times 5 \times 5 = 125$

D3 You *should* have been able to make a cube from 64 and 27 multilink cubes but *not* with 16 cubes.

D4 $1 = 1^3$ first odd number
$3 + 5 = 2^3$ next two odd numbers
$7 + 9 + 11 = 3^3$ next three odd numbers
$13 + 15 + 17 + 19 = 64 = 4^3$ next four odd numbers
$21 + 23 + 25 + 27 + 29 = 125 = 5^3$
and so on . . .

Talking Point

If the total number of cubes can be made by the same number multiplied by itself three times you can make a larger cube.
For example, with 64 cubes, $4 \times 4 \times 4 = 64$ so you can make a larger cube, but not with 16.

D5 Yes, Amy is correct because $13 \times 13 \times 13$ (or 13^3) is 2197.

D6 We are looking for a number which gives an answer of 4913 when it is cubed.

Number	20	15	18	17
(Number)³	8000	3375	5832	4913

D7 With a little practice you should be able to recognise all the whole number cube roots less than 100. The cube root of 64 is 4 ($4^3 = 64$).

E1 $1^3 + 3^3 + 6^3 = 1 + 27 + 216 = 244$
$3^3 + 4^3 + 5^3 = 27 + 64 + 125 = 216 = 6^3$
Both are true.

E2 The number 1 has the same square root as its cube root. It is 1!

E3 We need only to look at all the digits cubed to see if any whole number cubed ends in a two. Why is this?
$0^3 = 0$, $1^3 = 1$, $2^3 = 8$, $3^3 = 27$, $4^3 = 64$, $5^3 = 125$, $6^3 = 216$, $7^3 = 343$, $8^3 = 512$, $9^3 = 729$.
Some cubed whole numbers do end in a 2. Look at the list above. Are there any digits which the cube of a whole number cannot end in?

E4 The cube root of 100 ($\sqrt[3]{100}$) is approximately 4·641589, which is less than 5. Why do we say approximately?

E5 Can you make a cube from 1 000 000 cubes with none left over?
What this question is really asking is 'is the cube root of 1 000 000 a whole number?' The cube root of 1 000 000 is 100. Check that this is exact:
$100 \times 100 \times 100 = 1\,000\,000$.

E6 Is the cube of a number always greater than the square of the number? What about 0·1, 0·9, etc?

F1 This is your own investigation, but you might find this table useful.

number	$\sqrt{}$number	number	$\sqrt{}$number	halfway number	halfway $\sqrt{}$number
25	5	36	6	30·5	5·5227
36	6	49	7	42·5	6·5192
49	7	64	8	56·5	7·5166
64	8	81	9	72·5	8·5147

F2 Numbers which appear in the last digit or digits when they are squared are:
$5^2 = 25$, $76^2 = 5776$

F3 Try the trick on a friend. You can find the original numbers by adding one and subtracting one from the calculator display.

F4 Here are some number squares of odd numbers.

```
 1  3  5      1  3  5  7       1  3  5  7  9
 7  9 11      9 11 13 15      11 13 15 17 19
13 15 17     17 19 21 23      21 23 25 27 29
             25 27 29 31      31 33 35 37 39
                              41 43 45 47 49
```

For each grid, numbers on each diagonal give: 27 (3 by 3 grid), 64 (4 by 4 grid), 125 (5 by 5 grid). So the connection is that the diagonal totals are the size of the square cubed ($27 = 3^3$, $64 = 4^3$ and $125 = 5^3$).

F5 Here is a triangle of odd numbers.

```
                1
              3   5
           7    9    11
        13   15   17   19
     21   23   25   27   29
   31   33   35   37   39   41
```

There are several different patterns you should be able to see.
- The squares of the odd numbers $1^2 = 1$, $3^2 = 9$, $5^2 = 25$ and so on
- One less than the square of the even numbers 3, 15, 35, . . .
- $1 = 1^3$, $3 + 5 = 2^3$, $7 + 9 + 11 = 3^3$, $13 + 15 + 17 + 19 = 4^3$, . . .
- Some you discovered yourselves?

Try One of these

(1) The sum of all the numbers in each grid is 81 (3 by 3 grid), 256 (4 by 4 grid) and 625 (5 by 5 grid).
$81 = 3 \times 3 \times 3 \times 3$ (or 9^2)
$256 = 4 \times 4 \times 4 \times 4$ (or 16^2)
$625 = 5 \times 5 \times 5 \times 5$ (or 25^2)
What would you expect the sum of the numbers in the 6 by 6 grid to be? Check your answer.

(2) $1^3 = 1$ $\qquad = 1^2$
$1^3 + 2^3 = 9$ $\qquad = 3^2$
$1^3 + 2^3 + 3^3 = 36$ $\qquad = 6^2$
$1^3 + 2^3 + 3^3 + 4^3 = 100$ $\qquad = 10^2$
$1^3 + 2^3 + 3^3 + 4^3 + 5^3 = 225 = 15^2$
What's special about the numbers 1, 3, 6, 10, 15 . . .?

(3) What did you find out about number squares and triangles like these?

```
 2  4  6      2  4  6  8       2  4  6  8 10
 8 10 12     10 12 14 16      12 14 16 18 20
14 16 18     18 20 22 24      22 24 26 28 30
             26 28 30 32      32 34 36 38 40
                              42 44 46 48 50
```

$$
\begin{array}{ccccccc}
& & & 2 & & & \\
& & 4 & & 6 & & \\
& 8 & & 10 & & 12 & \\
14 & & 16 & & 18 & & 20 \\
22 & 24 & & 26 & & 28 & & 30 \\
32 & & 34 & & 36 & & 38 & & 40 & & 42 \\
44 & & 46 & & 48 & & 50 & & 52 & & 54 & & 56
\end{array}
$$

You should find some similar results to the squares and triangles of odd numbers. In what ways were your results different from before?

(4) You should have found that this sequence repeats itself. Do you always end up going around the loop no matter what the starting number is?

(5) You don't always end up going round the same loop. Does the sequence you get always repeat in the end?

Primes and factors

The **sum** of (say) 2, 6 and 8 is $2 + 6 + 8 = 16$.
The **product** of (say) 5, 2 and 3 is $5 \times 2 \times 3 = 30$.

A1 (a) The sum of 4 and 5 is $4 + 5 = 9$.
(b) $1 + 3 + 7 = 11$ (c) $1 + 10 + 2 = 13$

A2 (a) The product of 4 and 5 is 4×5 which is 20.
(b) $1 \times 3 \times 7 = 21$ (c) $1 \times 10 \times 2 = 20$

A3

If you pick any two or three numbers, their product will always be larger than their sum.

$1 + 10 + 2 = 13$ and $1 \times 10 \times 2 = 20$

$4 + 5 = 9$ and $4 \times 5 = 20$

$1 + 3 + 7 = 11$ and $1 \times 3 \times 7 = 21$

What about: $1 + 1$ and 1×1?
$0 \cdot 1 + 5$ and $0 \cdot 1 \times 5$,
and negative numbers?

A4 Looking for two numbers whose product is equal to their sum is quite hard.
If only whole numbers are allowed there are two pairs 2, 2 and 0, 0!
If you are allowed negative numbers and

decimals there are lots of possibilities. Here are a few of them:
$^{-}1 \cdot 5, 0 \cdot 6;$ $3, 1 \cdot 5;$ $1 \cdot 4, 3 \cdot 5;$ $^{-}1, 0 \cdot 5$.
Check them for yourselves.

B1 (a) The factors of 6 are 1, 2, 3 and 6.
(These all divide exactly into 6.)
(b) The factors of 24 are 1, 2, 3, 4, 6, 8, 12 and 24.
(c) The factors of 16 are 1, 2, 4, 8 and 16.
(d) The factors of 30 are 1, 2, 3, 5, 6, 10, 15 and 30.

Remember 1 and the number itself are always factors of the number.

B2 (a) The number 81 has these factors:
1, 3, 9, 27, 81 – it does have 5 factors.
(b) 510 cannot possibly be a factor of 223 307. At the very least if 510 is a factor of a number that number must end in a zero. Why?
(c) Not *all* numbers have at least 2 factors. Remember 1 which has only one.

Talking Points

Even numbers always have two as a factor and odd numbers never do. Why?
If 4 is a factor of a number, then the number is divisible by 4 so it must also have a factor 2.
You can always tell if a number has 5 as a factor – it ends in a 5 or 0. Why?

This short BASIC program will give the factors of any number n you input.
But it will not work for largish numbers. Why? If you know BASIC, improve the program – there are at least two ways you could do this.

```
10 INPUT "What numbers do you want
to find the factors of "; n
20 FOR number = 1 to n
30 IF n/number = INT (n/number)
THEN PRINT number
40 NEXT n
```

B3 Here is the completed worksheet W11.

1	2	3	4	5	6	7	8	9	10	11	12	13	14	15
			95											
			94											
			93											
			91											
			87											
			86											
			85											
	97		82											
	89		77											
	83		74											
	79		69											
	73		65											
	71		62											
	67		58											
	61		57											
	59		55											
	53		51		99									
	47		46		98									
	43		39		92									
	41		38		76									
	37		35		75									
	31		34		68									
	29		33		63		88							
	23		27		52		78							
	19		26		50		70							
	17		22		45		66							
	13		21		44		56							
	11		15		32		54				96			
	7	49	14		28		42				90			
	5	25	10		20		40				84			
	3	9	8	81	18		30	100	80		72			
1	2	4	6	16	12	64	24	36	48		60			

Number of factors

(a) The next number in column 3 is 121.

(b) The numbers in column 3 are the squares of the numbers in column 2. Do you think that this pattern will continue?

(c) 128 comes in column 8 and 236 is in column 9. Can you find a number which comes in column 11?

B4 ▲ Check column 2 of the completed worksheet W11.
Remember a prime number only has two factors (itself and 1).

B5 *Any whole number is either prime or the sum of two primes.*
This is not true. It does not work for these numbers 27, 35, 51, 57, etc. Once you have shown a rule does not work even once, you have proved that the rule is false. (To be a true rule it must work *all* the time.)
In fact *any even number is the sum of at most two prime numbers* has never been found not to work.

B6 Here are some of the multiples of six:
6, 12, 18, 24, 30, 36, 42, 48, . . .
One added to each of these gives:
7, 13, 19, <u>25</u>, 31, 37, 43, <u>49</u> . . .
The rule take any multiple of six and add one does not always give a prime number (all the underlined numbers are square numbers). What if we leave out square numbers – will the rest be prime? Investigate if you have time.

B7 If a number is a multiple of 4 it must also be a multiple of 2. Why?
A number which is a multiple of 6 is also a multiple of 2 and 3. Why?
The smallest multiple of 2, 3, 4, 5 and 6 is 60.

C1 ▲
(a) The factor-pairs of 64 are 1 and 64 ($1 \times 64 = 64$), 2 and 32 ($2 \times 32 = 64$), 4 and 16 ($4 \times 16 = 64$), 8 and 8 ($8 \times 8 = 64$).
(b) The factor-pairs of 19 are just 1 and 19, (19 is prime).

C2 There is not much point carrying on a factor tree when you reach a prime number, because a prime number has one set of factor-pairs, one and the number itself.

C3 If you multiply the end numbers of a factor tree the answer will be the starting number.
For this:

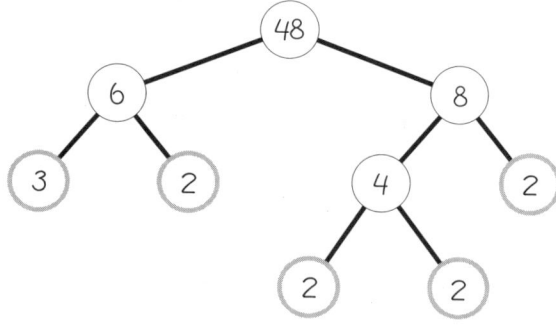

multiplying the end numbers gives
$3 \times 2 \times 2 \times 2 \times 2 = 48.$

C4 Here is different factor tree for 48.

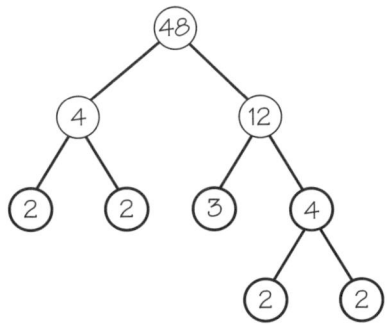

(b) $2 \times 2 \times 3 \times 2 \times 2 = 48$

C5 The bottom prime numbers multiplied together always give the starting number.

C6 Tell your teacher what you found.
▲

D1

A	B	C	D	E	F
1	2	3	4	5	6
7	8	9	10	11	12
13	14	15	16	17	18
19	20	21	22	23	24
25	26	27	28	29	30
31	32	33	34	35	36

(a) You would not expect a prime number in F, because the numbers in this row are all multiples of 6.

(b) There will be no primes (apart from 2) in B because these are all even numbers. Column D are also all even so there will never be any primes in this column.

D2

A	B	C	D	E
1	2	3	4	5
6	7	8	9	10
11	12	13	14	15
16	17	18	19	20

We can only be sure that, apart from 5, there will not be any primes in column E. Why?

Option

A	B	C	D	E	F	G
1	2	3	4	5	6	7
8	9	10	11	12	13	14
15	16	17	18	19	20	21
22	23	24	25	26	27	28

Can you see any patterns of primes (or non-primes) here?
                or here?

A	B	C	D	E	F	G	H	I
1	2	3	4	5	6	7	8	9
10	11	12	13	14	15	16	17	18
19	20	21	22	23	24	25	26	27
28	29	30	31	32	33	34	35	36

Talking Points

- Any number can be split up into prime factors. How does this help?
- If we are checking a number for "primeness", the largest number we need to check is round about the square root of that number.
  So to check 61 it would be 7.
  Check it yourselves. Why would it be silly to try 8, 9, etc.
  Talk through these again if you are still not sure.

D3 (a) For 461 the largest number you would need to check for divisibility is round about 21. Why? 461 is prime.

(b) 463 is prime. The numbers to test would be 3, 5, 7, 11, 13, 17 and 19.

(c) 467 is also prime. The only numbers you need to have used to check for divisibility were 3, 5, 7, 11, 13, 17 and 19.

D4 Your own words. (Look back in the booklet to check them.)

Patterns and sequences

Tables

A1 Take 1456 away from this year.

A2 From the table, $5 \times 43 = 215$.

A3 The six times table is:
$$1 \times 6 = 6$$
$$2 \times 6 = 12$$
$$3 \times 6 = 18$$
$$4 \times 6 = 24$$
$$5 \times 6 = 30$$
$$6 \times 6 = 36$$
$$7 \times 6 = 42$$
$$8 \times 6 = 48$$
$$9 \times 6 = 54$$
$$10 \times 6 = 60$$
$$11 \times 6 = 66$$
$$12 \times 6 = 72$$

The last digit in the answers gives the pattern: 6, 2, 8, 4, 0, 6, 2, 8, 4, 0

A4
(a) These numbers: 6, 0, 4, 8, 2, 6, 0 probably came from the last digit in the answers for the four times table.

(b) The last digit in the answers to the three times table makes this pattern 6, 9, 2, 5, 8, 1, 4, 7, 0, 3, 6

(c) 6, 8, 0, 2, 4, 6, 8, 0 is the pattern made by the last digit in the answers to the two times table.

A5 Try these tables: 4 times, 14 times, 24 times . . . (you may need a calculator!)

A6 Which tables did you find the hardest?

B1 ▲
(a) $5 + 8 = 13$ (b) $6 + 9 = 15$
(c) $3 + 8 = 11$ (d) $5 + 9 = 14$
(e) $2 + 3 = 5$
Did you find it quicker to use the table or your head?

B2 Here is an example. If you know that $3 + 6$ is 9, then $9 - 3$ is 6 and $9 - 6$ is 3, and so on.

B3 The number in the table is the side number subtracted from the top number. For example, $5 - 3$ is 2.

1	2	3	4	5	6	7	8	9	0
0	1	2	3	4	5	6	7	8	1
	0	1	2	3	4	5	6	7	2
		0	1	[2]	3	4	5	6	3
			0	1	2	3	4	5	4
				0	1	2	3	4	5
					0	1	2	3	6
.						0	1	2	7
							0	1	8
								0	9

Parts (a) and (d) of **B2** can be done using this table.

C1

1	[2]	3	[4]	[5]	6	7	8	9	[10]	11	12
[2]	4	6	<u>8</u>	10	12	14	16	18	20	22	24
3	6	9	<u>12</u>	15	18	21	24	27	30	33	36
[4]	8	12	<u>16</u>	20	24	28	32	36	40	44	48
[5]	<u>10</u>	<u>15</u>	<u>20</u>	25	30	35	40	45	50	55	60
6	12	18	24	30	36	42	48	54	60	66	72
7	14	21	28	35	42	49	56	63	70	77	84
8	16	24	32	40	48	56	64	72	80	88	96
9	18	27	36	45	54	63	72	81	90	99	108
[10]	20	30	40	50	60	70	80	90	100	110	120
11	22	33	44	55	66	77	88	99	110	121	132
12	24	36	48	60	72	84	96	108	120	132	144

C2 Look carefully at the table.
$$20 = 10 \times 2 \text{ (or } 2 \times 10\text{)}$$
$$20 = 4 \times 5 \text{ (or } 5 \times 4\text{)}$$

C3 ▲ Say you wanted to find $24 \div 4$.
- Find 4 along the side of the table.
- Now follow along from the 4 till you reach 24.
- Read up to the top of the table – the answer is 6.

C4 You should have got these right, but just in case!
(a) $3 \times 2 = 6$ (b) $6 \times 3 = 18$
(c) $8 \times 4 = 32$ (d) $7 \times 7 = 49$

C5 If, for example, you take 9×3 as 3×9 the multiplication tables are the same.

Try this

This is *one* way to find 14×8.
Split 14 into $7 + 7$, so 14×8 is the
same as $7 \times 8 + 7 \times 8$
Think carefully about this.
From the tables $7 \times 8 = 56$ so that
$14 \times 8 = 56 + 56 = 112$.
You may know of some other methods
yourselves.

C6 (a) 153 (b) 91 (c) 168
▲

Investigate

Here are some ways to think about it.
$19 \times 12 = 10 \times 12 + 9 \times 12$
$19 \times 12 = 19 \times 2 \times 2 \times 3$
$19 \times 12 = 19 \times 6 + 19 \times 6$
$19 \times 12 = 20 \times 12 - 12$

D1 (a)
▲

The answers
in the ten
times table
always end
in a zero.

This is true. To prove
it you only have to
look at all the times
tables from 1 to 9 up to
ten times.
Why is this? Tell your
teacher what you both
think.

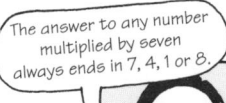

The answer to any number
multiplied by seven
always ends in 7, 4, 1 or 8.

(b)

What about 7×5?

(c)

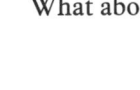

There is always a pattern
in the last digit for all the
times tables.

This is true.

(d) What about the five times
table?

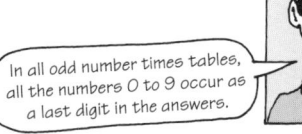

In all odd number times tables,
all the numbers 0 to 9 occur as
a last digit in the answers.

D2 Did you find that in all the even times
tables the last digit is always even?
Did you discover anything else?

D3 (a) $176 \times 5 = 881$, must be wrong, the
answer should end in a zero. Why?
(b) $671 \times 10 = 6710$, looks right, it ends in
a zero.
(c) $567 \times 12 = 6803$, must be wrong, it
should end in a 4.
(d) $13 \times 11 = 145$, is wrong because it
should end in a 3.

Challenge

One way to check a division is to do the
multiplication with your answer.
For example, to check $12 \div 4$ is 3, work
out 3×4, the answer should be 12.
Here is another example, if $60 \div 4$ is 15,
then 15×4 should be 60.
Did you find any other checks of your
own for division?

D4 (a)

$1 \times 1 = 1$	$7 \times 7 = 49$
$2 \times 2 = 4$	$8 \times 8 = 64$
$3 \times 3 = 9$	$9 \times 9 = 81$
$4 \times 4 = 16$	$10 \times 10 = 100$
$5 \times 5 = 25$	$11 \times 11 = 121$
$6 \times 6 = 36$	$12 \times 12 = 144$

(b) 114×114 must end in a 6.
(c) 36×36 must end in a six.

E1 (A) matches with (f), (B) matches with (c),
(C) matches with (d), (D) matches with (a),
(E) matches with (e), (F) matches with (b)
and (G) matches with (g).

F1 ▲

	Number	Result of casting out the nines
(a)	5678	8
(b)	234	9
(c)	46	1 (4 + 6 = 10, 1 + 0 = 1)
(d)	975	3
(e)	756	9

234 and 756 are divisible by 9.

F2 What did you both notice?

You should have found that the remainder you get when you divide a number by 9 is always the same as the answer when you cast out nines from that number.

F3 The answers in box 4 and box 6 are always the same.

F4 ▲

	Number	Result of casting out the nines
(a)	157	4
	17	8
	4 × 8 = 32	5
	2669	5
(b)	342	9
	43	7
	9 × 7 = 63	9
	14704	7

(a) 157 × 17 is probably 2669 – can you say why?

(b) 342 × 43 is not 14704 – how can you tell?

F5 What did you find out?

If you did not try this question, see if casting out nines works for this calculation 5152 ÷ 14 = 368.

Number	Result of casting out the nines
368	8
5152	4
14	5

Can you find a way to work out the result of casting out nines from 4 ÷ 5?

16

Option

Here is a program which will print out all the tables up to 12 × 12.
You may be able to make it print a little neater than this does.

```
10 FOR table = 1 TO 12
20 FOR number = 1 TO 12
30 PRINT number; "X"; table; "=";
number*table
40 NEXT number
50 NEXT table
```

G1 ▲

(a) 3 *in* 4 = 12 — *in* probably meant multiply.

(b) 4 *mo* 18 = 22 — *mo* looks as if it means to 'add'.

(c) 10 *mi* 4 = 6 — *mi* means to subtract – probably short for *minus*.

(d) 4 *p* 2 = 6 — *p* means add.

(e) 5 *et* 5 = 10 — *et* stands for add.

(f) 12 ÷ 4 = 8 — for this calculation ÷ means subtract! A few hundred years ago a symbol might mean several different things.

(g) 4 * 2 = 8 — * means to multiply on computer key boards. Why do you think this is so?

Challenge

It looks as if the numbers on the right could be the numbers on the left multiplied by themselves or squared. What do you think ⟪ stands for?

What's next?

A1 Here are the completed patterns.

(a)

(b)

(c)

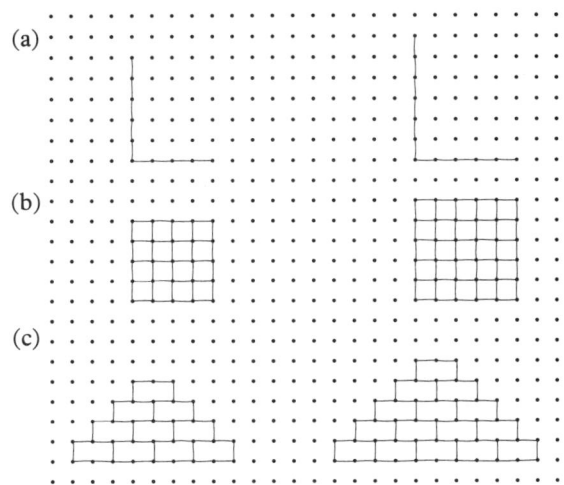

A2 Your own patterns.

A3

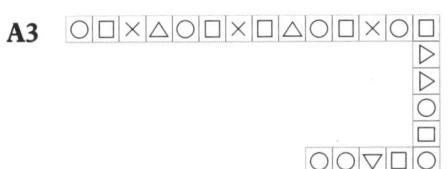

A [○○] B [○□×△○]

C [□×□△] D [○□×]

E [△△○□] F [○□△]

G [○□]

The pieces fit in this order, starting from the left. B C D G E F A

A4 Your own puzzles.

B1

$4 \xrightarrow{+2} 6 \xrightarrow{+2} 8 \xrightarrow{+2} 10 \xrightarrow{+2} ?$

The number after 10 will be 12.

B2 ▲
  (a) The next two numbers in the sequence 1, 4, 7, 10, 13, 16 are 19 and 22.
  (b) To get from one **term** to the next you need a '+3' number machine.

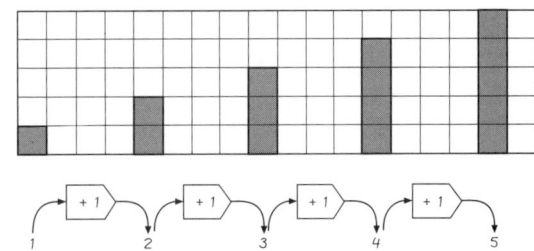

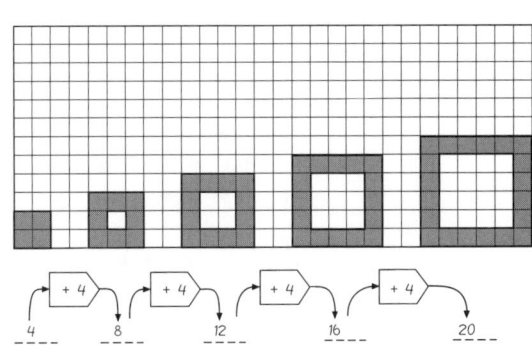

C1 The next two **terms** in the sequence 30, 27, 24, 21 are 18 and 15.

C2 The next term in this sequence

$4 \xrightarrow{×2} \xrightarrow{+3} 11 \xrightarrow{×2} \xrightarrow{+3} 25$

is 53 ($25 × 2 = 50$, and $50 + 3 = 53$).

C3 (a) The next three terms in the sequence 7, 15, 23 are 31, 39 and 47.
  (b) The number machine rule for the sequence is [+8 ⟩

C4 This number machine rule fits the sequence 1, 3. [×2 ⟩—[+1 ⟩

C5 ▲ There are many of different number machine rules which give the sequence starting 2, 4. Here are just a few.

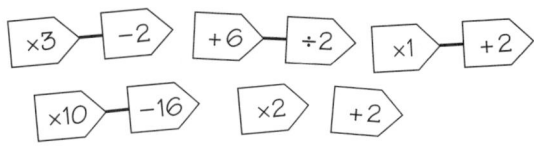

Show your teacher any different ones which you have found.

D1 (a)

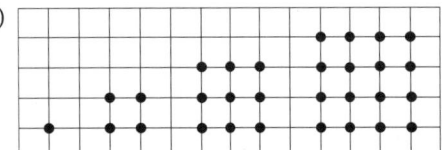

The numbers in this sequence are called square numbers.
It goes 1 (1×1) 4 (2×2) 9 (3×3)
16 (4×4) 25 (5×5) 36 (6×6)
49 (7×7) and so on.

(b)

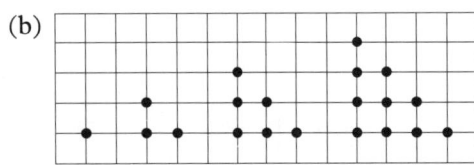

This sequence is called the sequence of triangle numbers.
Can you see why?
The sequence goes 1, 3, 6, 10, 15, 21, 28 and so on.

(c)

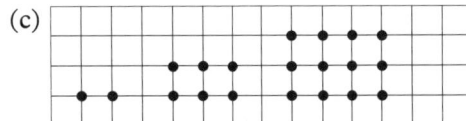

These numbers could be called the sequence of rectangle numbers.
The sequence goes 2, 6, 12, 20, 30, 42 and so on.
Can you see anything special about this sequence?

(d)

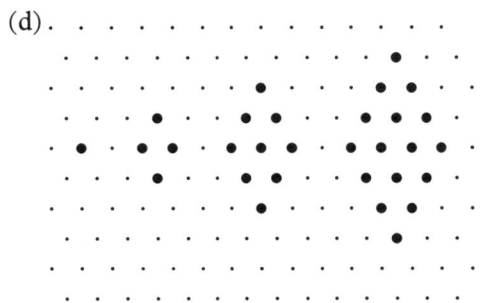

You could call this the sequence of diamond numbers.
It goes 1 4 9 16 25 36 49 ...
It's the square numbers.
Does it remind you of any other number sequence?

(e)

This might be called a sequence of trapezium numbers. You may have thought of a better name yourselves!
The sequence goes 1, 5, 12, 22, 35, 51, ...

D2 Your own answers.

D3 (a) 1 = 1
▲   1 + 3 = 4
       1 + 3 + 5 = 9
       1 + 3 + 5 + 7 = 16
       1 + 3 + 5 + 7 + 9 = 25
These are the square number – they are made by adding together odd numbers.
(b) 1 = 1
     1 + 2 = 3
     1 + 2 + 3 = 6
     1 + 2 + 3 + 4 = 10
The answers to these give the sequence of triangle numbers.
(c)   1 + 3 = 4
      3 + 6 = 9
      6 + 10 = 16
      10 + 15 = 25
These are the square numbers.
They are made by adding together two triangle numbers in order. Take a look at the sequence of triangle numbers: 1, 3, 6, 10, 15, 21, 28 ...

Challenge

Try rearranging the cubes, for example:

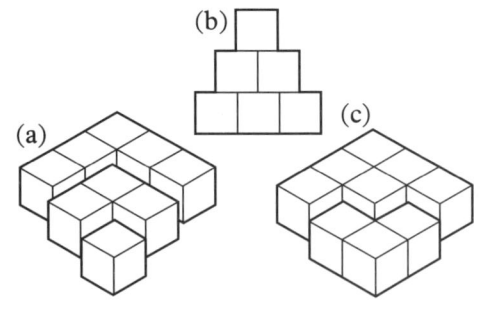

(a)   (b)   (c)

18

Number patterns and puzzles 1

A1 (a)

```
      4   (6 + 1)
  6   1   7
(4 + 1) 5
```

(b)

```
      3   (9 + 4)
  9   4   13
(3 + 4) 7
```

(c)

```
      5   (6 + 4)
  6   4   10
(5 + 4) 9
```

A2 (a)

```
      6
  3  10  13
(13 – 10) 16
   (6 + 10)
```

(b)

```
      9
  2   6   8
(8 – 6) 15
   (9 + 6)
```

(c)

```
(18 – 6)  4
      12  6  18
(10 – 4) 10
```

A3 (a)

```
      6  (3 + 10)
  3  10  13
(6 + 10) 16
```

(b)

```
      3  (11 – 3)
  10  8  18
(18 – 8) 11
```

(c)

```
      1  (14 – 13)
  6  13  19
(19 – 6) 14
```

(d)

```
(23 – 18) 27
      5  18  23
(27 + 18) 45
```

(e)

```
         23
  8   8  16
(16 – 8) 31
```

(f)

```
(43 – 25) 18  (28 + 25)
      28  25  53
         43
```

A4 There are two ways of doing this question. You can put the numbers this way round … or this way round

```
      2                8
  8   5  13        2   5   7
      7                13
```

(But are the two ways really different?)

A5 You could put your answers 'the other way round' to these, like the ones above.

(a)

```
      1
  12  6  18
      7
```

(b)

```
      3
  9   7  16
      10
```

(c)

```
      2
  18  6  24
      8
```

(d)

```
      9
  15  16  31
      25
```

B1 (a)

7	8	15
5	3	8
12	11	23

(b)

8	6	14
3	3	6
11	9	20

(c)

6	3	9
11	10	21
17	13	30

(d)

0	8	8
3	9	12
3	17	20

(e)

11	8	19
3	15	18
14	23	37

(f)

0	16	16
14	0	14
14	16	30

(g)

8	2	10
21	11	32
29	13	42

(h)

6	7	13
16	2	18
22	9	31

(i)

0	0	0
0	0	0
0	0	0

C1 Check your own numbers carefully.

0	3	6	9	12
5	8	11	14	17
10	13	16	19	22
15	18	21	24	27
20	23	26	29	32

The closer you look at the number square the more patterns you see! For example:
0, 8, 16, 24, 32, …
12, 14, 16, 18, 20, …

C2

0	3	6	9	12
5	8	11	14	17
10	13	16		
15	18			

The rule is **add 8**.

C3

0	3	6	9	12
5	8	11	14	17
10	13	16		
15	18			

The rule is **subtract 2**.

C4

0	3	6	9	12
5	8	11	14	17
10	13	16	19	22

The rule is **add 6**.

C5 The rule is **subtract 10**.

C6 **Subtract 16** is the rule here.

C7 The rule is **subtract 7**.

C8 Check your own numbers carefully.

1	2	4	8
3	6	12	24
9	18	36	72
27	54	108	216

C9 The rule is **multiply by 6**.

C10

1	2	4	8
3	6	12	
9	18		
27			

Multiply by 4 is the rule.

C11

1	2	4	8
3	6	12	
9	18		
27			

Divide by 2 is the rule.

C12 The rule is **divide by 3**.

C13 The rule is **divide by 9**.

C14

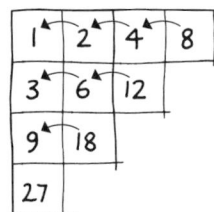

(a) Multiply by 3.

(b) Multiply by 2.

(c) Multiply by 6.

D1

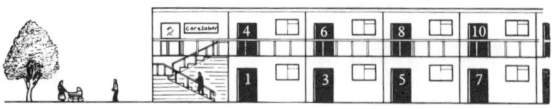

(a) These numbers are all on the top floor: 30, 46 and 50.

(b) 11 is below 14.   (c) 34 is above 31.

(d) The number on the flat below is always 3 less.

(e) You add 3 to the bottom flat number to get the top flat number.

D2

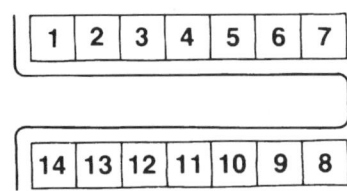

(a) These are all tall houses: 16, 32, 40 and 60.

(b) These house numbers are next door to tall houses: 17, 29 and 47.

D3 (a) 36   (b) 43   (c) 69

D4

1	2	3	4	5	6	7

14	13	12	11	10	9	8

The two numbers opposite each other always add up to 15.

D5

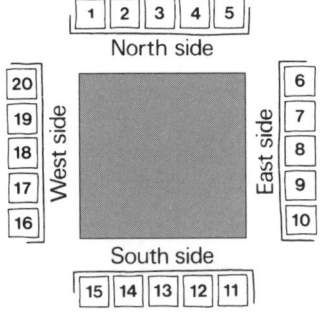

(a) North and South numbers add up to 16.

(b) East and West house numbers opposite each other add up to 26.

E1 Check your own numbers.
Remember the rules are:
left-hand number × middle number = right-hand number
top number × middle number = bottom number

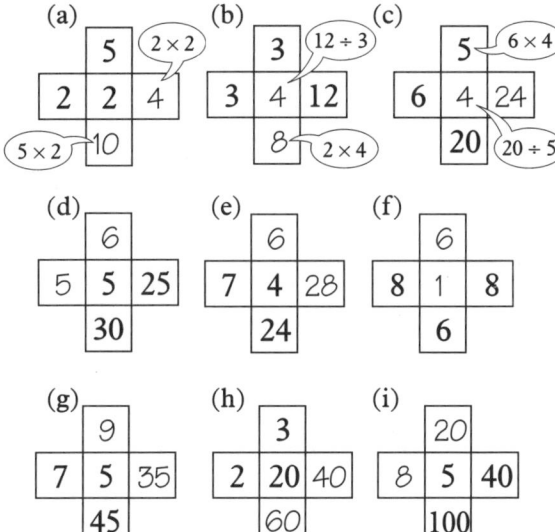

(a)
5 (2 × 2)
2 2 4
(5 × 2) 10

(b)
3 (12 ÷ 3)
3 4 12
8 (2 × 4)

(c)
5 (6 × 4)
6 4 24
20 (20 ÷ 5)

(d)
6
5 5 25
30

(e)
6
7 4 28
24

(f)
6
8 1 8
6

(g)
9
7 5 35
45

(h)
3
2 20 40
60

(i)
20
8 5 40
100

E2 There are two ways to fit the numbers
4, 2, 10, 5, 20 into a multiplication cross.
These are:

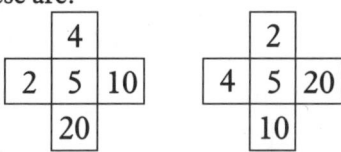

4
2 5 10
20

2
4 5 20
10

(Is there a connection between the two
multiplication crosses?)

E3 Each of these answers could be the other
way round.

(a)
2
3 6 18
12

(b)
2
7 4 28
8

(c)
3
6 5 30
15

(d)
6
10 4 40
24

(e)
2
3 0 0
0

E4
2
3 2 6
4

2
5 3 15
6

5
4 2 8
10

3
2 4 8
12

This rule fits all the multiplication grids:
Left-hand number × bottom number =
top number × right-hand number.

F1

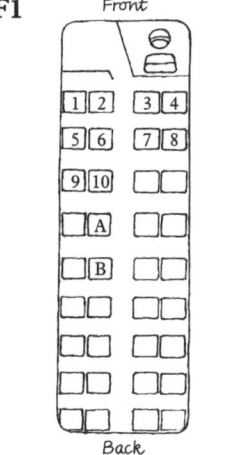

(a) Seat 14 is the seat A.
(b) Seat B is number 18.

F2 (a) Seat 16 is one behind seat 12.
(b) Seat 19 is one behind seat 15.

F3 (a) Seat 26 is one behind seat 22
(b) Seat 31 is one behind seat 27.

F4 The next seat behind is always 4 more.

F5

David is sitting in seat 20.

F6 (a) Seat 24   (b) Seat 14

F7 14 adults can go.

G1 Coach A, British and Danish
(25 + 9 = 34)
Coach B, French and German
(29 + 13 = 42)
Coach C, Spanish (19)

G2 7 cats (They have 7 × 4 legs, Mr Smith has
2, giving 30 legs in all.)

G3 11 cm (53 − 9 = 44, 44 ÷ 4 = 11)

G4 5 miles ($8 \times 2 = 16, 66 - 16 = 50$, $50 \div 5 = 10, 10 \div 2 = 5$. Don't forget all the journeys are 'there and back'!)

G5 22 miles (half the walk is $6 + 5 = 11$ miles, so the whole walk is $2 \times 11 = 22$).

G6 14 people (The same number of people must have got on the bus as were on it in the first place – think about it!)

G7 Show your teacher.

Sequences

A1 (a) These are the first ten numbers in the sequence of even numbers:
2 4 6 8 10 12 14 16 18 20
(b) The 50th number in the sequence of even numbers is 100.
(c) The millionth number in the sequence of even numbers is 2 000 000.
How did you work this out?
(d) 100 is the 50th number in the sequence of even numbers.
How could you convince someone that this is true?

A2
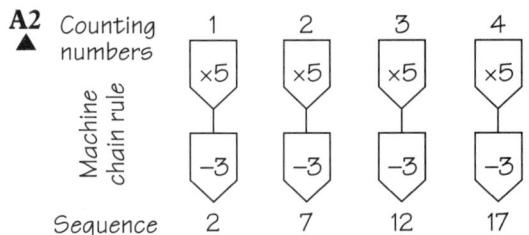

(a) The 10th term in the sequence is 47 ($10 \times 5 = 50, 50 - 3 = 47$).
(b) The 20th term is 97 ($20 \times 5 = 100$, $100 - 3 = 97$).
(c) 42 is the 9th term. Did you use what you learnt in *Doing and undoing*?

A3 (a) The first ten terms which this machine chain makes are:
3 7 11 15 19 23 27 31 35 39
(b) The 50th term is 199 ($50 \times 4 = 200$, $200 - 1 = 199$).

A4
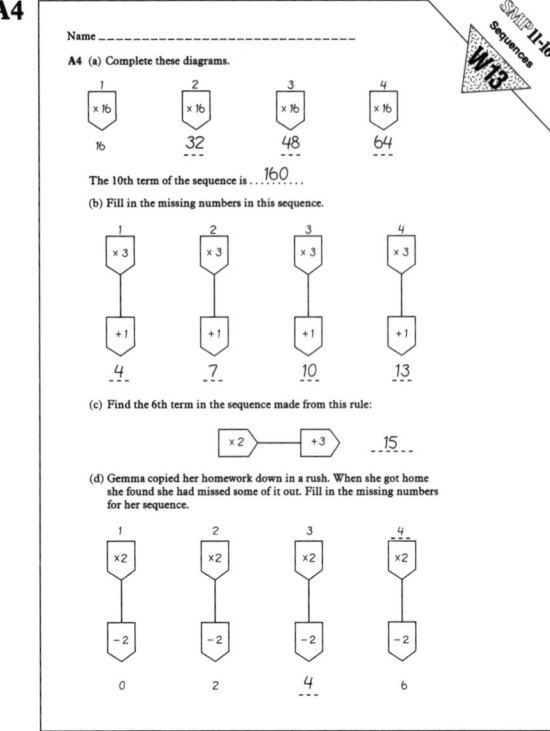

A5
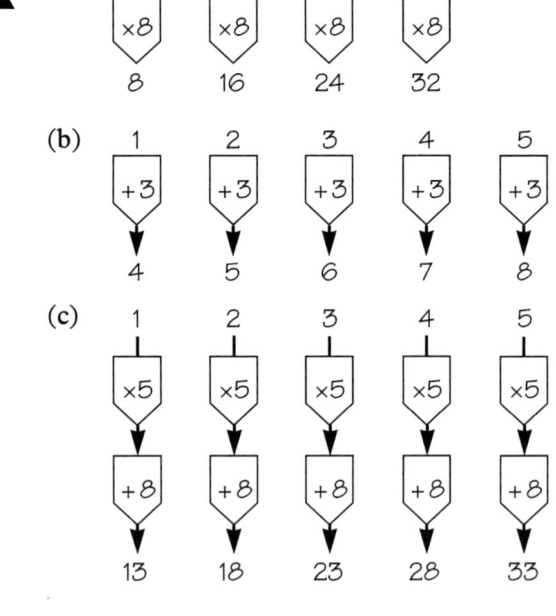

22

(d)

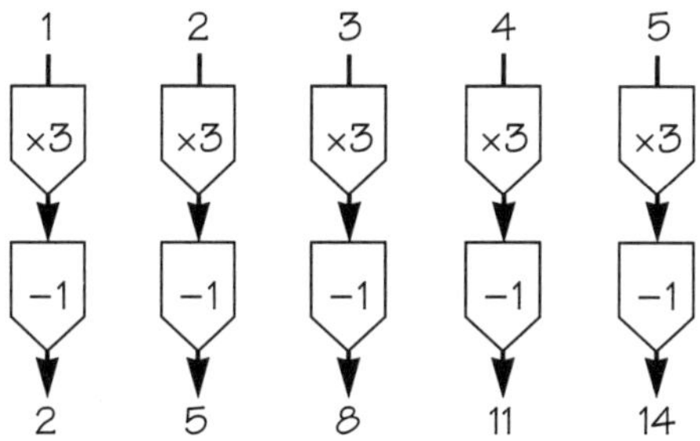

A6 This rule will give the sequence of odd numbers 1, 3, 5, 7, . . .
If you are not sure, check if for yourselves.

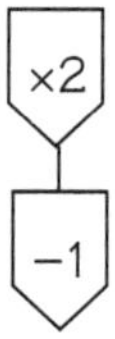

Try One of these

(1) Probably the easiest way to do this is by 'trial and improvement' just by looking closely at the machine chain. The 50th term is 200 – 1 = 199, the 51st term is 204 – 1 = 203.

(2) If you did this, show your teacher your program.

B1 (a) The sixth term in the sequence 1 4 9 16 25 is 36 (6 × 6 = 36).
(b) The 6th term is 6 × 6, the 7th term is 7 × 7, and so on.
(c) The 10th term is 10 × 10 which is 100.
(d) What we are looking for is a number which gives 81 when multiplied by itself. 9 × 9 = 81, so 81 is the 9th term in the sequence.

B2 (a) The 7th square number is 49.
(b) The 21st square number is 441.

B3 These are all square numbers:
16 36 49 100.
You may find it useful to know all the square numbers up to 100.

B4 The 31st square number is 961, the 32nd is 1024. So the 32nd is nearer to 1000?

Challenge

Hint. experiment with the key marked [√] if you have one on your calculator.
The last digit in a square number must be one of these
0 1 4 5 6 or 9.
So 15 129 could be a square number and 678 963 is not.

B5

Square number	1	4	9	16	25	36	49	64	81
Digital root	1	4	9	7	7	9	4	1	9

Square number	100	121	144	169	196	225	256	289	324
Digital root	1	4	9	7	7	9	4	1	9

Investigate

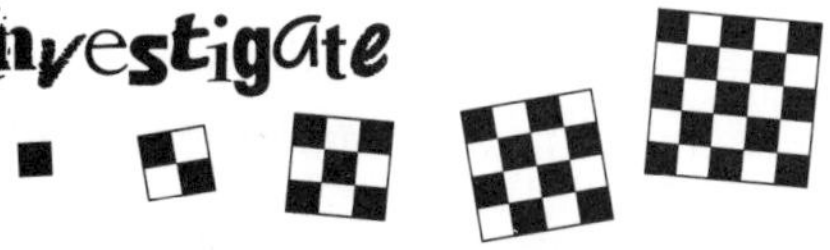

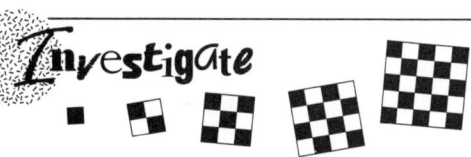

The number of black tiles is shown in the table. You will need to extend it. What do you notice?

Total number of tiles	1	4	9	16	25	36	49
Number of black tiles	1	2	5	8	13	18	25

C1 This is the dot pattern for the 6th triangle number. It is 21.

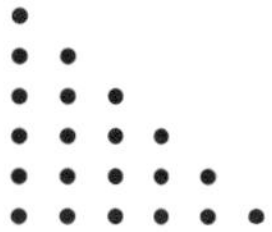

C2 ▲ Here is a list of the first ten triangle numbers.
1 3 6 10 15 21 28 36 45 55
In this list of triangle numbers, 1 and 36 are also square numbers.

C3 The fifth triangle number is 15. The first five counting numbers added together also give 15. Can you see from the pattern of dots why this works for all triangle numbers?

C4 If you add the fourth and the fifth triangle numbers together you get $10 + 15 = 25$ which is a square number. Did you get square numbers for your examples?
Can you fit together the patterns of dots to see why this happens?

C5

Triangle number	1	3	6	10	15	21	28	36	45
Digital root	1	3	6	1	6	3	1	9	9

The digital root of a triangle number is always 1, 3, 6 or 9.

Challenge

The first 20 triangle numbers are 1, 3, 6, 10, 15, 21, 28, 36, 45, 55, 66, 78, 91, 105, 120, 136, 153, 171, 190, 210.
How does the sequence carry on?

D1 Here are the next two pentagon numbers.

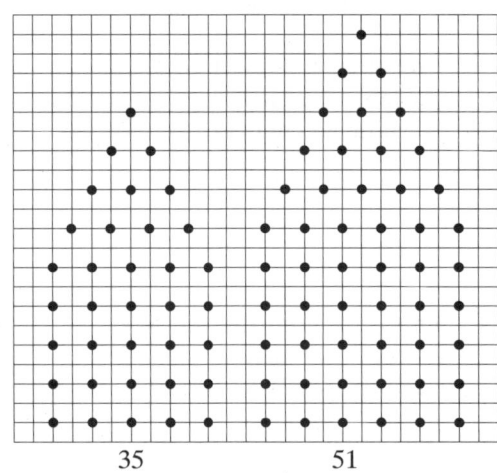

35   51

D2 Here are the first six pentagon numbers
    1   5   12   22   35   51
The first five pentagon numbers are all less than 50.

D3 This is a bit of a trick question! 1 is a square, triangle and pentagon number.

D4

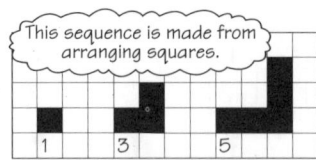

This sequence is made from arranging squares.

Here is the sequence of gnomon numbers.

Term	1	2	3	4	5	6	7	8	9
Gnomon number	1	3	5	7	9	11	13	15	17

D5 Some things you may have noticed were:
- all the numbers are odd
- the difference between each number is two. Could you explain why these are true by looking at how the gnomon numbers are made?
- what else did *you* find out?

D6 (a) The first 3 gnomon numbers added together $(1 + 3 + 5)$ give 9.
(b) The first 4 added give $1 + 3 + 5 + 7$ which is 16.
(c) The answers to (a) and (b) are square numbers. Look carefully at how you can fit together the diagrams of the first three gnomon numbers. Can you use it to explain this result?
(d) The first 106 gnomon numbers add up to $106 \times 106 = 11236$.

D7 The sequence formed by taking the first gnomon number from the first square number, the second gnomon number from the second square number and so on, is the square numbers.
    0   1   4   9   16   25   36

D8 In fact Winston and Eli made a mistake. Their sequence should have been the sequence of triangle numbers. Show your own report to your teacher.

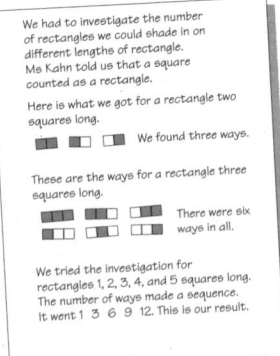

We had to investigate the number of rectangles we could shade in on different lengths of rectangle. Ms Kahn told us that a square counted as a rectangle.

Here is what we got for a rectangle two squares long.

We found three ways.

These are the ways for a rectangle three squares long.

There were six ways in all.

We tried the investigation for rectangles 1, 2, 3, 4, and 5 squares long. The number of ways made a sequence. It went 1 3 6 9 12. This is our result.

Try this

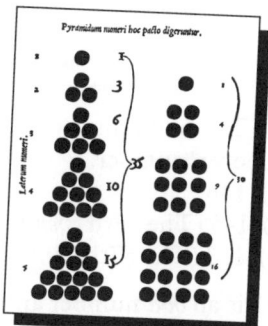

For the triangular pattern, the balls in each layer are triangle numbers.
In the case of the square pattern, the number in each layer is a square number.
For the triangular pattern the 'pyramid' numbers are:

1   1 + 3 = **4**   1 + 3 + 6 = **10**
1 + 3 + 6 + 10 = **20**
1 + 3 + 6 + 10 + 15 = **35**

For the square arrangement the sequence is different.

1   1 + 4 = **5**   1 + 4 + 9 = **14**
1 + 4 + 9 + 16 = **30**
1 + 4 + 9 + 16 + 25 = **55**

D9 Here is one way to make dot shapes from the sequence 2, 6, 12, 20, . . .

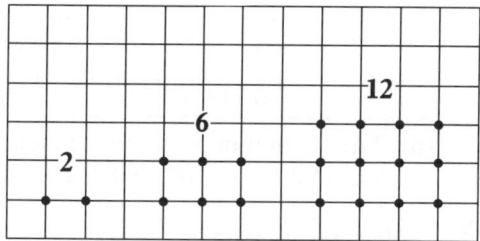

(a) One name for the sequence could be the rectangle numbers.
(b) The first 7 terms of the sequence are 2, 6, 12, 20, 30, 42, 56.
(c) If you divide each term by two, you get this sequence:
   1, 3, 6, 10, 15, 21 . . .
   the triangle numbers.

D10 The sequence of cube numbers is:
   1   8   27   64   125 . . .
One number which is a cube, square and triangle number is 1. There are some numbers which are both cube and square numbers. For example, $64 = 4^3$ and $64 = 8^2$. Can you find some more?
Only 1 is a cube and triangle number.

E (a) This is the sequence you get:
   5   17   29   41
   53   **65**   (!)

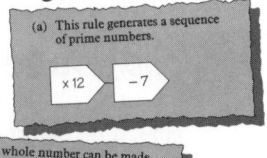

(a) This rule generates a sequence of prime numbers.

× 12   − 7

(b) This is true.

(b) Any whole number can be made up from no more than three triangle numbers. For example, 8 = 6 + 1 + 1.

(c) This is also true.

(c) One plus eight times any triangle number always gives a square number.

F1

Sequence name	Terms					
	1	2	3	4	5	6
Triangle	1	3	6	10	15	21
Square	1	4	9	16	25	36
Pentagon	1	5	12	22	35	51
Hexagon	1	6	15	28	45	66
Septagon	1	7	18	34	55	81
Octagon	1	8	21	40	65	96
Nonagon	1	9	24	46	75	111

Show your teacher the patterns you both found.

F2 (a) A septagon has seven sides. You can work this out from the table.
(b) If the third term in Audrey's shape is 30 then the shape has 11 sides.

F3 You should find a pattern in the last digits of the triangle numbers, but it only repeats after 20 terms. Can you find a pattern within the first 20 terms? What patterns did you find in the other sequences?

F4 (a) The next row would be
   1   9   36   84   126   126   84   36   9   1
(b) When you add the numbers in each row you get the sequence
   1   2   4   8   16   32   64 . . .
This sequence is made by multiplying 2 by itself: $4 = 2 \times 2$, $8 = 2 \times 2 \times 2$, and so on.
(c) Can you find the counting numbers and the triangle numbers? Did you find any other number sequences?

F5 The next row in this pattern is:
1 6 21 50 90 126 141 126 90 50 21 6 1
Can you find the counting numbers and
the triangle numbers in the triangle?
Did you find any other number patterns?
What numbers do you get when you add
together the numbers in a row?

Challenges

(1) The pattern odd, odd, even, even
does carry on with the triangle
numbers. What is the pattern for
the square numbers? Did you find
any other patterns like this?

(2) 1 is both a pentagon and a triangle
number. Did you find any others?

(3) Look at the answer to **B5** for the
digital roots of the square numbers
and at the answer to **C5** for the
digital roots of the triangle
numbers. Is there a pattern in the
digital roots of the cube numbers?
Did you try any other sequences?

(4) A square number is a number
multiplied by itself and so it must
have a factor other than 1 and itself
(unless it actually is 1). This means
that it cannot be prime.

(5) Many people have looked for a
pattern in the sequence of prime
numbers, but they have been
unable to find one. Have you?

(6) Your own answer.

Number sequences

A1 (a) The 15th even number is 30.
(b) 104 is the 52nd even number.
(c) The 124th even number is 248.

A2 (a) The 10th odd number is 19.
(b) 23 is the 12th odd number.
(c) 39   (d) 79

A3

	1st	2nd	3rd	4th	5th	6th	7th	8th	9th	10th	11th
Even numbers	2	4	6	8	10	12	14	16	18	20	22
Odd numbers	1	3	5	7	9	11	13	15	17	19	21

(a) The odd number is 1 less than the
even number.
(b) 146
(c) 145 (The 73rd *even* number is 146.)
(d) 133 (The 67th *even* number is 134.)

A4 237 is an *odd* number and 238
(237 + 1 = 238) is the 119th *even* number
(238 ÷ 2 = 119). So 237 is the 119th *odd*
number.

A5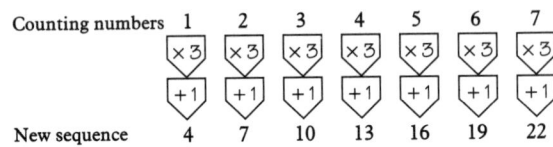

The 20th number in the sequence is 61.
(20 × 3 = 60, 60 + 1 = 61)

A6 (a)

Counting number	1	2	3	4	5	6	7	8	9	10
Sequence	2	5	8	11	14	17	20	23	26	29

(b) The 40th number in the sequence is
119, (40 × 3 = 120, 120 − 1 = 119).

A7 (a)

Counting number	1	2	3	4	5	6	7	8	9	10
Sequence	7	12	17	22	27	32	37	42	47	52

(b) The 100th number in the sequence is
502. (100 × 5 = 500, 500 + 2 = 502)

A8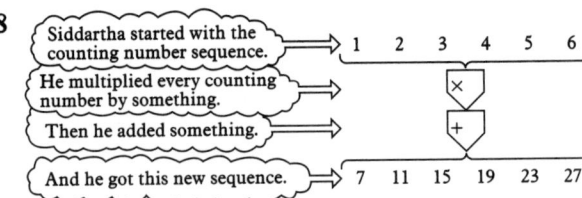

Siddartha started with the counting number sequence.
He multiplied every counting number by something.
Then he added something.
And he got this new sequence.

He multiplied by 4, then added 3.

A9 (a) × 5 (multiply by five)
(b) × 5, + 3 (multiply by five then add three)
(c) × 4, + 1   (d) × 3, − 1
(e) × 10, + 3   (f) × 4, − 3

A10 (a) 175 (35×5)

(b) 178 ($35 \times 5 = 175,\ 175 + 3 = 178$)

(c) 141 ($35 \times 4 = 140,\ 140 + 1 = 141$)

(d) 104 ($35 \times 3 = 105,\ 105 - 1 = 104$)

(e) 353 ($35 \times 10 = 350,\ 350 + 3 = 353$)

(f) 137 ($35 \times 4 = 140,\ 140 - 3 = 137$)

A11 (a) Divide by 2

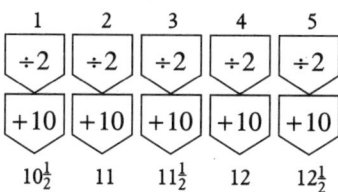

(b) Divide by 2, then add 10. (Or add 20 and divide by 2.)

B1 (a) This is the next dot pattern after 25.

● ● ● ● ● ●
● ● ● ● ● ●
● ● ● ● ● ●
● ● ● ● ● ●
● ● ● ● ● ●
● ● ● ● ● ●

(b) 36 (6×6)   (c) 49 (7×7)

B2 (a) The 9th square number is 81 (9×9).

(b) The 17th square number is 289. (You do 17×17.)

B3 (a), (b) and (c)

1 4 9 16 25 36 49 64 81 100 121 144 169 196
+3 +5 +7 +9 +11 +13 +15 +17 +19 +21 +23 +25 +27

What is special about the bottom numbers?

B4 (a)

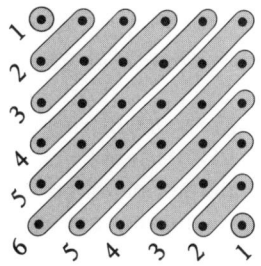

$36 = 1 + 2 + 3 + 4 + 5 + 6 + 5 + 4 + 3 + 2 + 1$

(b) $64 = 1 + 2 + 3 + 4 + 5 + 6 + 7 + 8 + 7 + 6 + 5 + 4 + 3 + 2 + 1$

(c) For 5×5 the middle number was 5 for 6×6 it was 6.
So for 23×23 the middle number should be 23.
How can you be sure?

B5

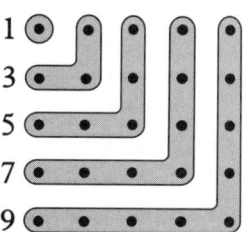

This drawing shows that 25 can be written as $1 + 3 + 5 + 7 + \mathbf{9}$.

(a) For 49,
$1 + 3 + 5 + 7 + 9 + 11 + 13 = 49$.

(b) Look at the diagram. For 25 which is 5×5 the last number is $2 \times 5 - 1 = 9$.
For 49 (7×7) the last number is $2 \times 7 - 1 = 13$.
So for 100, which is 10×10, the last number is 19.
How could you convince someone this is true?

(c) For 529 (23×23) the last number is 45 ($46 - 1$).

B6

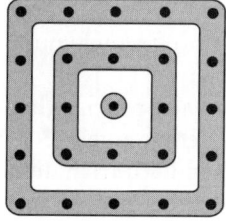

Here is another way of splitting up 25.

27

(a) If you add another 'ring' you get
1 + 8 + 16 + 24 = **49**
(which is 7 × 7!) Can you explain from the diagrams why the outer 'ring' must have 24 dots in it?

(b) The number of dots in each 'ring' makes a number pattern.
The pattern is clearer if you look at this table.

Total number of dots	1 (1 × 1)	9 (3 × 3)	25 (5 × 5)	49 (7 × 7)	81 (9 × 9)
Number of dots in the outer 'ring'	1	8	16	24	32

For 529 (529 = 23 × 23, and 23 is the 12th odd number) there will be 88 dots in the outer 'ring'.

C1 This is the 6th triangle number.

C2

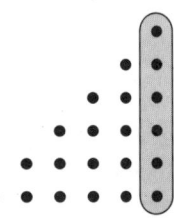

You need to add six dots to the 5th triangle to make the 6th triangle number.

C3 You need to add 7 dots to the 6th triangle to get the 7th triangle.

C4 (a) 24 dots added to the 23rd triangle give the 24th triangle.
(b) The 24th triangle number is the 23rd triangle number plus 24.
It is 276 + 24 = 300
(c) Adding 25 to 300 will give the 25th triangle number.

C5 ▲ (a) The first two triangle numbers added together give 4 (1 + 3).
(b) The 2nd and 3rd triangle numbers added together give 9 (3 + 6).
(c) The 3rd and 4th triangle numbers added together give 16 (6 + 10).
(d) 25
(e) The answers are square numbers.

C6 Here is one way to split this dot diagram into two triangle number diagrams.

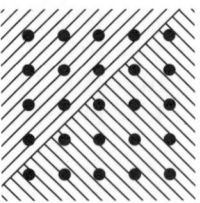

C7 The 5th and 6th triangle numbers add up to 36 (15 + 21 = 36).

C8 (a) 4th triangle number + 5th triangle number = 5 × 5
5th triangle number + 6th triangle number = 6 × 6
(b) 8th triangle number + 9th triangle number = 81 (9 × 9)
(c) 16th triangle number + 17th triangle number = 289 (17 × 17)
(d) The difference between the 17th and 16th triangle numbers is 17.
(e) The 16th and 17th triangle numbers add up to 289.
They have a difference of 17.
The two triangle numbers are 136 and 153.
Could you explain to someone how you worked this out?

C9 Each layer starting from the top is a triangle number.
So in 6 layers there are
1 + 3 + 6 + 10 + 15 + 21 = 56 oranges.

D1

(a) The numbers in the top row are all square numbers.
(b) The 20th number in the top row will be 400 (20 × 20).
(c) The numbers in the first column are all one more than a square number.

(d) The 11th number in the first column is 101.

(e) The sequence formed by the numbers which go diagonally downwards from the corner is:

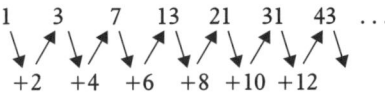

D2

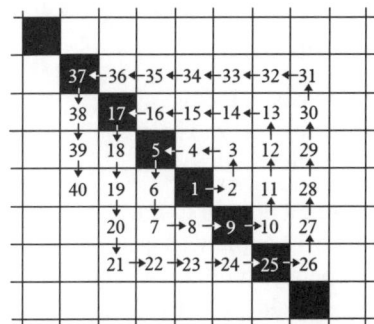

(a) The numbers in the top row are triangle numbers.

(b) The sequence which goes diagonally downwards from the corner goes 1, 5, 13, 25 then **41, 61, 85, 113, 145 ...** (The difference between the numbers goes in a sequence like this 4 8 12 16 20 24 28 32 ...)

D3

(a) The numbers in the black squares going downwards are
1, 9, 25, 49, 81, 121, 169, 225, ...
$(1 \times 1)\,(3 \times 3)\,(5 \times 5)\,(7 \times 7)\,(9 \times 9)$
$(11 \times 11)\,(13 \times 13)\,(15 \times 15)$

(b) The numbers in the black squares going upwards are
5, 17, 37, 65, 101, 145, 197, 257.
$(2 \times 2 + 1 = 5),\ (4 \times 4 + 1 = 17),$
$(6 \times 6 + 1 = 37),\ (8 \times 8 + 1 = 65)$ and so on.

D4 (a) The next five numbers in the grey squares going downwards are 43, 73, 111, 157, 211, ...

(b) The next five numbers in the grey squares going upwards are 57, 91, 133, 183, 241, ...

Multiplication and division patterns

In the 1993 reprint of this booklet, some of the prices are being changed to make them more realistic. If this reprint, or a later one, is being used, you need to look at the answers in square brackets for questions D1, D7 and D10.

A1
$3 \times 4 = 12$
$30 \times 4 = 120$
$300 \times 4 = 1200$
$300 \times 40 = 12\,000$
Can you see any patterns here?

A2 Check your answers to these.
(a) $70 \times 50 = 3500$
(b) $400 \times 700 = 280\,000$
(c) $80 \times 5000 = 400\,000$
(d) $30 \times 40\,000 = 1\,200\,000$
(e) $600 \times 500 = 300\,000$
(f) $30 \times 40 \times 50 = 60\,000$
(g) $800 \times 60 \times 20 = 960\,000$
(h) $400 \times 400 = 160\,000$

A3 If a person's heart beats 70 times in a minute then:
(a) in 1 hour it will beat 4200 times (60×70)
(b) in 200 hours it will beat 840 000 times $(200 \times 60 \times 70)$.

A4 There are 1 200 000 bolts in the storeroom $(30 \times 80 \times 500)$.

A5 There are roughly 120 000 words in the whole book $(300 \times 40 \times 10)$.

A6 (a) $80 \times ? = 48\,000$, so ? = 600 $(48\,000 \div 80)$
(b) $? \times 400 = 200\,000$, so ? = 500 $(200\,000 \div 400)$
(c) $? \times 5000 = 4\,000\,000$, so ? = 800
(d) $700 \times ? = 63\,000\,000$, so ? = 90\,000

Remember
'12 divided by 4' can be written as $12 \div 4$ or $\frac{12}{4}$.

B1   (a) $\frac{20}{4} = 20 \div 4 = 5$ (b) $\frac{18}{3} = 6 \ (18 \div 3)$

    (c) $\frac{72}{9} = 8 \ (72 \div 9 = 8)$ (d) $\frac{144}{2} = 72$

    (e) $\frac{57}{3} = 19$ (f) $\frac{108}{6} = 18$

C1   £24 is to be divided equally between 6 people.

    (a) Each person gets £4 (24 ÷ 6).

    (b) If there was twice as much money each person would get £8 (48 ÷ 6).

    (c) If there were twice as much money and twice as many people each person would still get £4 (48 ÷ 12 = 4).

C2   (a) £36 ÷ 12 = £3   (b) £18 ÷ 6 = £3
    (c) £36 × 5 = £180, 5 × 6 = 30,
    £180 ÷ 30 = £6
    (d) £3   (e) £3

C3   (a) Dividing 6 by $1\frac{1}{2}$ is the same as $\frac{6}{1\frac{1}{2}}$ which is the same as $\frac{12}{3}$ and this is 12 ÷ 3 = 4.

    (b) $\frac{14}{3\frac{1}{2}} = \frac{28}{7} = 28 \div 7 = 4$

    (c) $\frac{7\frac{1}{2}}{1\frac{1}{2}} = 15 \div 3 = 5$

    (d) $\frac{22\frac{1}{2}}{4\frac{1}{2}} = \frac{45}{9} = 5$

    (e) $\frac{27}{2\frac{1}{4}} = \frac{54}{4\frac{1}{2}} = \frac{108}{9} = 12$

    (f) $\frac{56\frac{1}{4}}{6\frac{1}{4}} = \frac{112\frac{1}{2}}{12\frac{1}{2}} = \frac{225}{25} = 225 \div 25 = 9$

C4   (a) 96 ÷ 16 = 6   (b) 144 ÷ 48 = 3
    (c) 84 ÷ 14 = 6   (d) 360 ÷ 24 = 15
    (e) 540 ÷ 18 = 30   (f) 2000 ÷ 25 = 80

C5   (a) $\frac{84}{3\frac{1}{2}} = \frac{168}{7} = 168 \div 7 = 24$

    (b) $\frac{405}{45} = \frac{810}{90} = 810 \div 90 = 9$

    (c) $\frac{180}{4\frac{1}{2}} = \frac{360}{9} = 40$

    (d) $\frac{192}{32} = \frac{96}{16} = \frac{48}{8} = 6$
(Does it matter in which order you do the divisions?)
    (e) $\frac{72}{2\frac{1}{4}} = \frac{144}{4\frac{1}{2}} = \frac{288}{9} = 32$

D1   (a) 4 cakes cost 31p
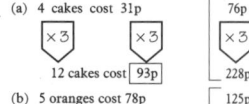 
12 cakes cost 93p   228p

    (b) 5 oranges cost 78p   125p
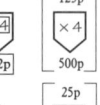 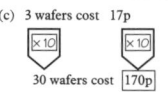
20 oranges cost 312p   500p

    (c) 3 wafers cost 17p   25p
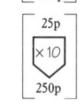 
30 wafers cost 170p   250p

    (d) 7 bottles hold 9 litres
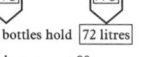
56 bottles hold 72 litres

    (e) 13 lemons cost 90p   £1·80
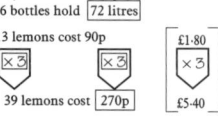 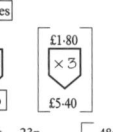
39 lemons cost 270p   £5·40

    (f) $7\frac{1}{2}$ ounces of pasta cost 23p   48p
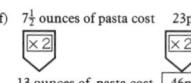 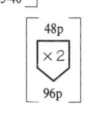
13 ounces of pasta cost 46p   96p

D2   (a) 5 pens cost 85p.

1 pen costs 17p
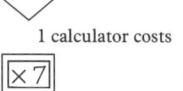
3 pens cost 51p

    (b) 3 calculators cost £36.
÷3   ÷3
1 calculator costs £12
×7   ×7
7 calculators cost £84

D3   8 cubic centimetres of metal weigh 56 grams.
    (a) 1 cubic centimetre weighs (56 ÷ 8) 7 grams.
    (b) 5 cubic centimetres weigh (7 × 5) 35 grams.

D4   6 tea-sets cost £210.
1 tea-set costs £35 (210 ÷ 6).
11 tea-sets cost £385 (11 × 35).

D5   9 radios cost £153.
1 radio costs £17 (153 ÷ 9).
So 16 radios will cost £272 (17 × 16).

D6   To take 50 people to Paris costs £3100.
To take 5 people will cost £310.
So 30 people will cost £1860.
(There are lots of different ways of solving this problem.)

D7   7 lb cost £2·66, so 1 lb costs £0·38 (or 38p).
So 5 lb will cost £1·90 (5 × 0·38).
[(7 lb cost £3·50, so 1 lb costs £0·50 (or 50p).
So 5 lb costs £2·50 (5 × £0·50).]

D8   8 plastic boxes cost £9·12.
1 box costs £1·14.
11 boxes will cost £12·54 (11 × 1·14).

D9

    (a) 49p per kg (343 ÷ 7)
    (b) 55p per kg (220 ÷ 4)

D10 (a) 18p per lb (54 ÷ 3)
    (b) 17p per lb (136 ÷ 8)
    [(a) 33p per lb (99 ÷ 3)
    (b) 31p per lb (248 ÷ 8)]

D11 Here is one way to solve the problem.
A piece of hardboard 2m by $1\frac{1}{2}$m
weighs 1·5 kg. This piece of wood has an
area of 3 square metres. In other words
3 square metres weighs 1·5 kg. So 1 square
metre weighs 0·5 kg. This means that a
piece of hardboard 5 m by 4 m (20 square
metres) will weigh 10 kg.

Number patterns and puzzles 2

A1 (a) To get from a black number to a green
    number you multiply by 4.
    (b) Green 28 goes with black 7.
    (c) Green 280 goes with black 70.
    (d) Black 12 goes with green 48.
    (e) ? = 60 (15 × 4)
    (f) ? = 8 (32 ÷ 4)
    (g) ? = 55 (220 ÷ 4)
    (h) ? = 88 (21 × 4 = 84, 84 + 4 = 88)
    (i) ? = 52 How did you work this out?
    (j) ? = 76 (72 + 4)

B1 (a)

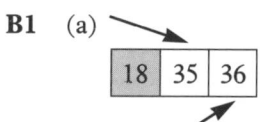

    (b) (35 = 2 × 18 − 1, 36 = 2 × 18)

B2 To find a right-hand number when you
know the left-hand one multiply it by 2.

B3 | 21 | | 42 | **B4** | 35 | | 70 |

B5 (a) The missing number is 64.
    (b) The missing number is 63.

B6 | 20 | ? | |

    ? = 39 (39 = 40 − 1)

B7 (a) | 17 | ? | (b) | 60 | ? | (c) | 43 | ? | |

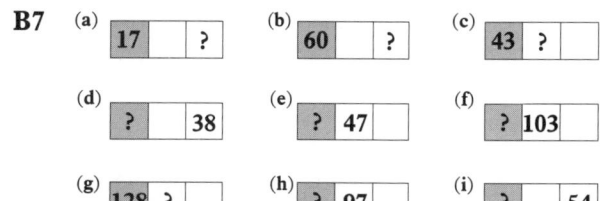

(d) | ? | 38 | (e) | ? | 47 | | (f) | ? | 103 | |

(g) | 128 | ? | | (h) | ? | 97 | | (i) | ? | | 54 |

Check your answers against these for the
numbers marked ?.
(a) 34   (b) 120   (c) 85   (d) 19
(e) 24   (f) 52   (g) 255   (h) 49
(i) 27

B8 (a) ? = 41
    (b) You can find ?
    by dividing 82 by 2.

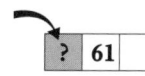

| ? | 82 |

B9 (a) ? = 31
    (b) To find ? you add 1
    onto 61, then divide by 2.

| ? | 61 | |

C1 The numbers on the bottom row can all be
divided exactly by 3.

C2 These numbers will all be in the bottom
row: 24, 27, 30, 60, 90, 120.

C3 The shaded (or coloured) tiles are all even
numbers.

C4 These numbers will all go onto shaded
(coloured tiles): 22, 28, 38, 52, 70, 90.

C5 The numbers on the coloured tiles on the
bottom row can all be divided exactly by 6.

C6 All these numbers can be on coloured tiles
on the bottom row:
24, 30, 42, 48, 60, 300.

Column → 1   2   3   4   5   6   7
numbers

1	4	7	10	13	16	19
2	5	8	11	14	17	20
3	6	9	12	15	18	21

C7 To get a bottom number you multiply the
column number by 3.

C8

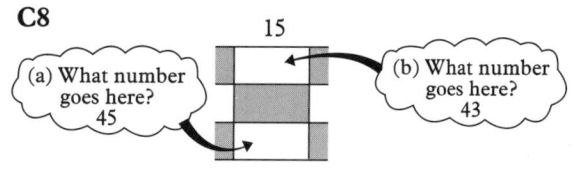

15

(a) What number goes here? 45

(b) What number goes here? 43

C9 To find the number on the top tile you multiply the column number by 3, then subtract 2.

C10

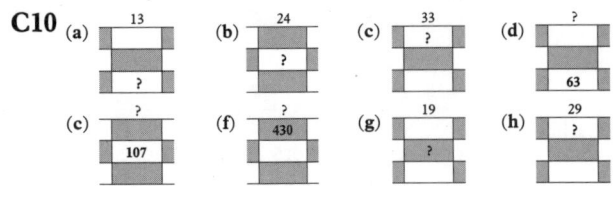

(a) 13 ?
(b) 24 ?
(c) 33 ?
(d) ? 63
(e) ? 107
(f) ? 430
(g) 19 ?
(h) 29 ?

(a) ? = 39 (b) ? = 71 (c) ? = 97
(d) ? = 21 (e) ? = 36 (f) ? = 144
(g) ? = 56 (h) ? = 85

D1 (a) ? = 80
(b) To get '80' you have to multiply 20 by 4.

D2 (a) ? = 121
(b) To find ? you multiply 30 by 4, then add 1.

D3 (a) ? = 12
(b) To find ? you subtract 1 from 49, then divide by 4.

D4 (a) 42 will be in the 7th grey square.
(b) 10 points to 61.
(c) 20 points to 121.
(d) 100 points to 601.

D5 These numbers are all on shaded (green) squares: 73, 121, 1201.

D6 You can work out which numbers are on shaded (green) squares by subtracting 1 and seeing whether what is left can be divided exactly by 6.

D7 (a) The arrow points to 122.
(b) You probably worked this out by multiplying by 4 then adding 2.

E1

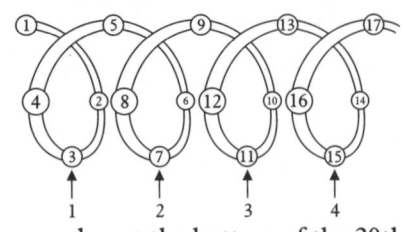

The number at the bottom of the 20th loop is 79.

E2 All these numbers are at the bottom of a loop: 23, 43, 47, 55.

E3 (a) The tenth number out from here is 30.
(b) The tenth number out on this line is 29.
(c) The 16th number on this line is 47.

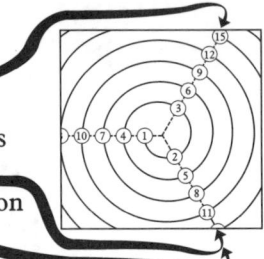